Science Pop Art
Biomes & Ecosystems

A Terminology Coloring Book

Illustrated and Written By

Mizzz Foster

Published by Fantastic Absurdities
Lake Elsinore, CA US
Email: MizzzFoster@gmail.com

ISBN: 978-1-7377913-3-1
Printed in the United States of America
Design by M. A. Foster
Cover Design by M. A. Foster

SCIENCE POP ART
BIOMES & ECOSYSTEMS

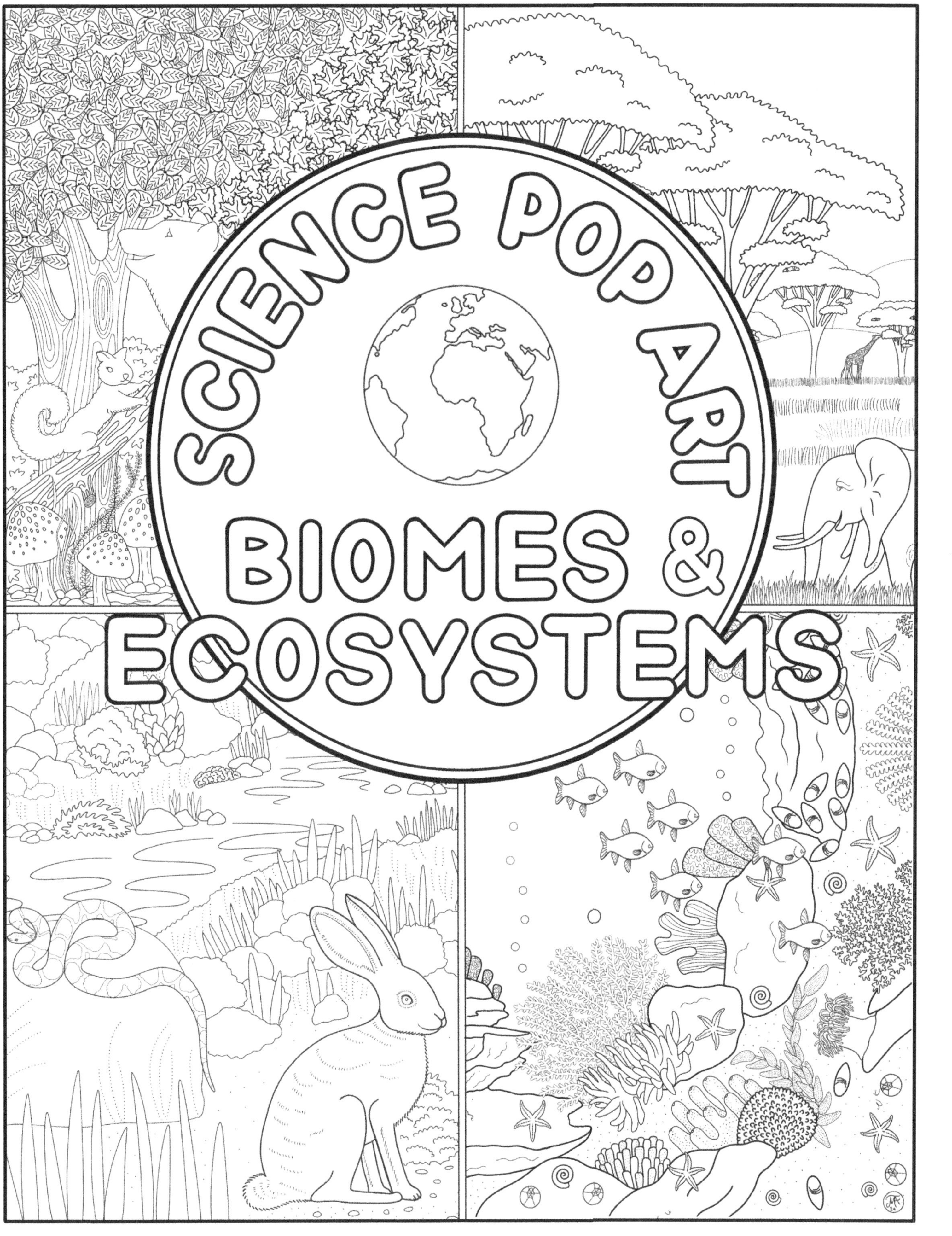

Table of Contents

Biome: A large region of the Earth characterized by specific climate conditions, plants, animals, fungus, protists, and bacteria. Biomes are defined by abiotic factors such as climate, soil type, and water availability, as well as biotic factors including the plants and animals adapted to the region. Earth has five major biomes; aquatic, grassland, forest, desert, and tundra. The aquatic biome is Earth's largest biome covering approximately 71% of Earth's surface.

BIOME

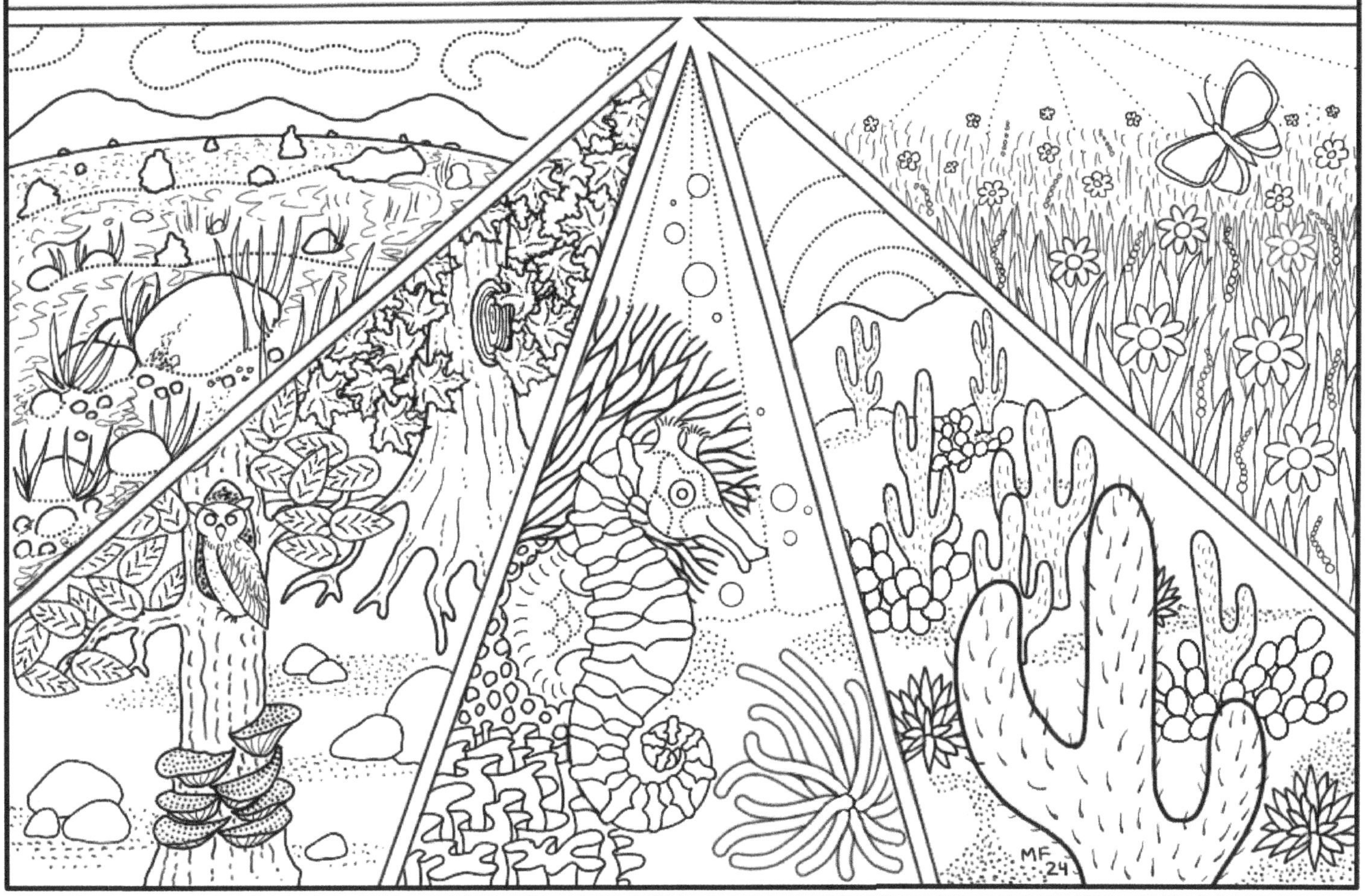

What is a biome?

What are the five major biomes?

Which is the largest biome on Earth?

What are the biotic elements of a biome?

What are the abiotic elements of a biome?

Ecosystem: A group of living organisms (biotic factors) interacting with one another and with their non-living environment (abiotic factors) in a specific area. These interactions create a balanced system where energy flows and matter cycles. Ecosystems can vary widely in size and complexity, from a lake or river to an expansive forest or ocean system. Each one has unique characteristics depending on its location and conditions. Animals and plants in each ecosystem evolved specific adaptations to help them thrive in each ecosystem.

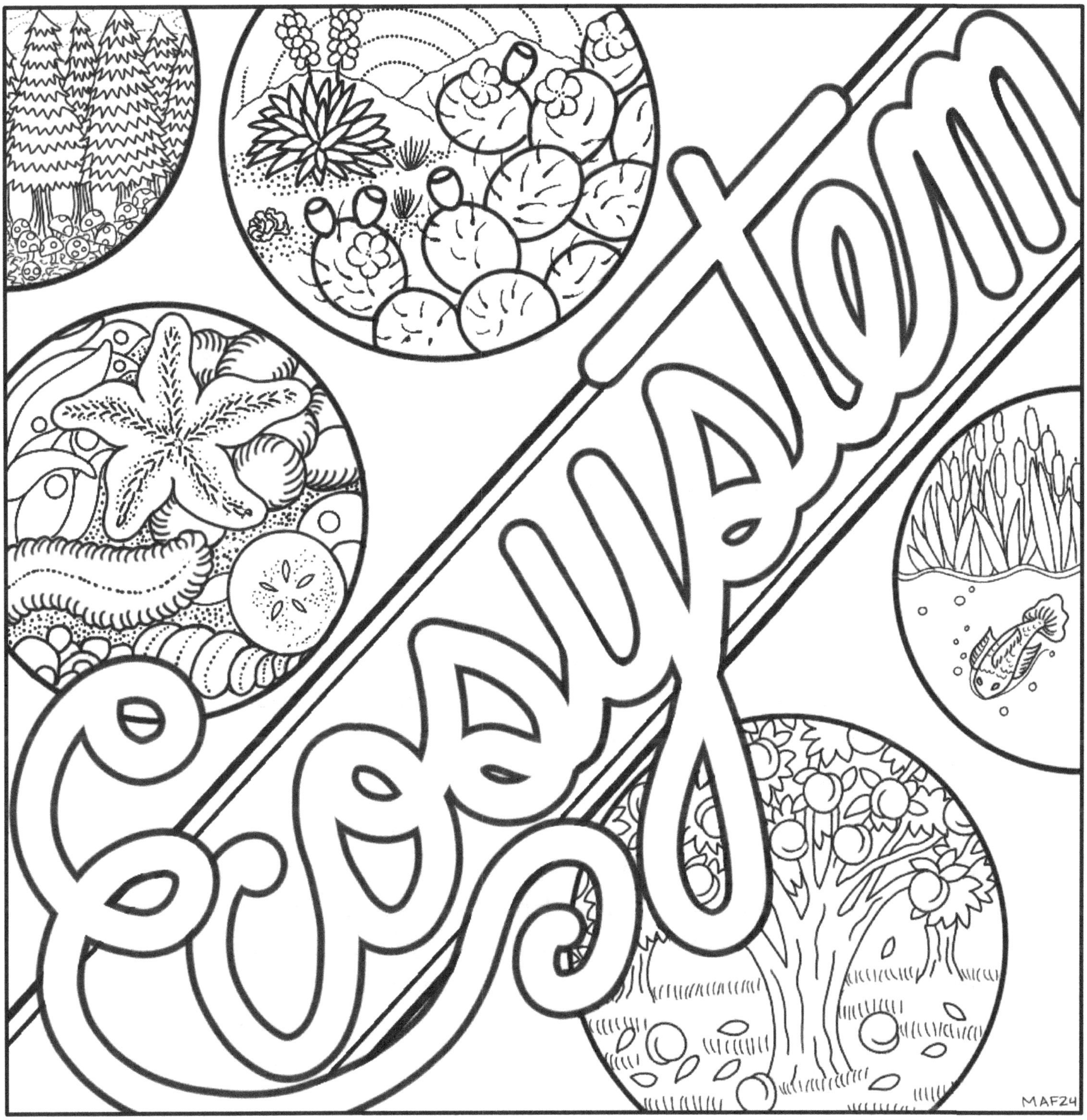

What is an ecosystem?

List six different ecosystems.

Aquatic Biome: The aquatic biome encompasses all water-based environments on Earth, covering 71% of the planet's surface. The defining feature of the aquatic biome is the presence of water, which shapes the environment and determines the types of organisms that can live in and around it.

The aquatic biome can be categorized by water; freshwater (e.g., lakes, rivers) or marine (e.g., oceans, coral reefs). Important abiotic factors of the aquatic biome are water salinity, light availability, temperature, dissolved oxygen, and nutrient levels.

The aquatic biome has producers such as algae, phytoplankton, and aquatic plants (e.g., seagrasses). The consumers are fish, amphibians, crustaceans, mollusks, and marine mammals. The decomposers are aquatic bacteria, fungi, and protists. The aquatic biome is important for humans because we get food in the form of fish and seaweed, but we also mine for oil, and use the water for transportation.

The aquatic biome is threatened by pollution, overfishing, climate change, habitat destruction, and invasive species. Aquatic pollution made up of chemicals, plastics, and waste harm water quality and aquatic life. Overfishing depletes fish populations which disrupts natural food chains. Climate change causes ocean warming, stronger storms, acidification, and rising sea levels. Habitat destruction from dredging, coral reef damage and wetland drainage reduce biodiversity. Invasive species disrupt native ecosystems and food webs by outcompeting native and keystone species.

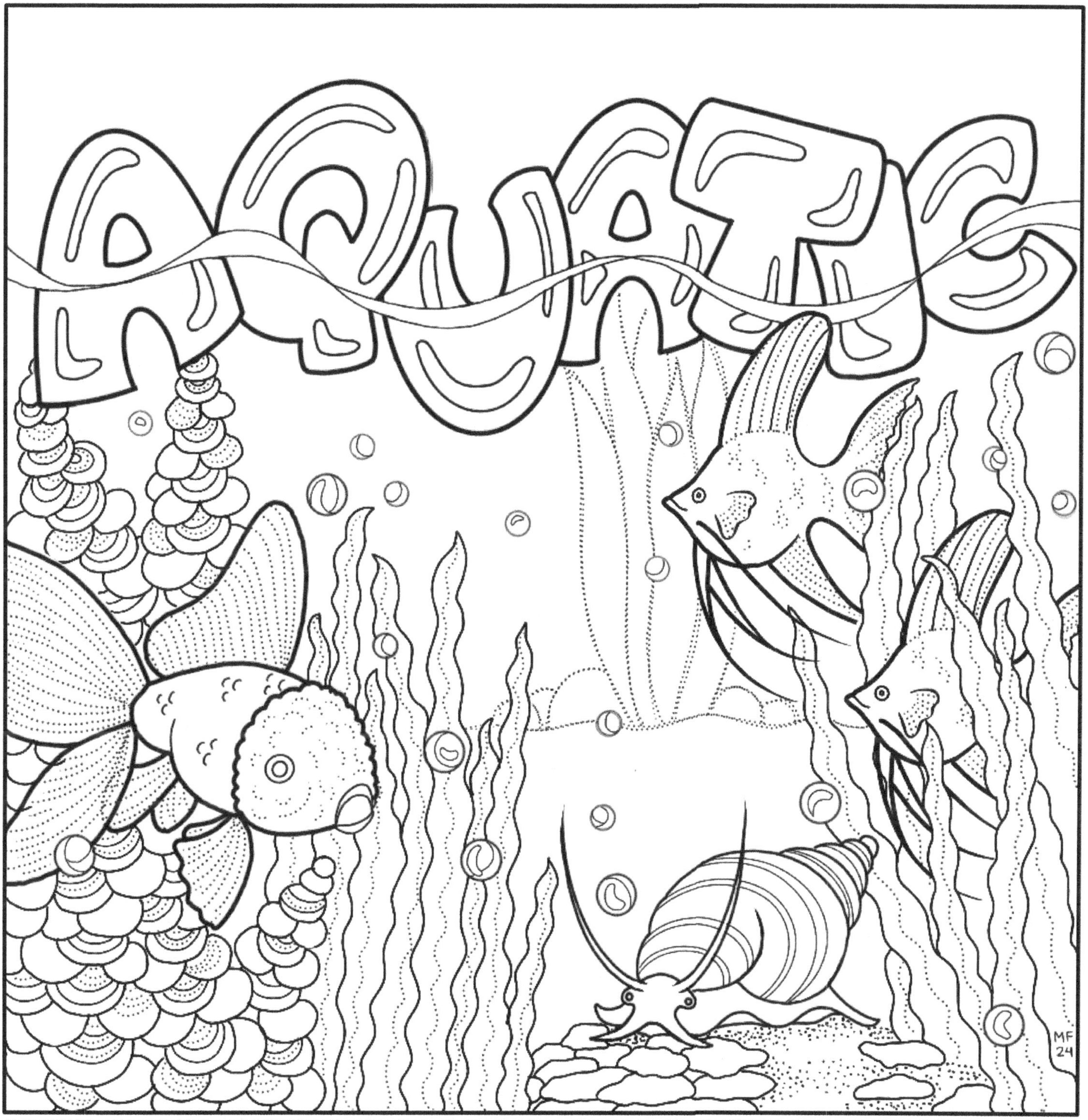

What are the two types of water found in the aquatic biome, and how do they differ?	What are two major threats to the health of the aquatic biome?
What are three abiotic factors that influence life in the aquatic biome?	How do invasive species damage the aquatic biome?

Marine Water Ecosystem: An aquatic ecosystem that includes all saltwater environments, such as oceans, seas, coral reefs, estuaries, and hydrothermal vents. Marine water is the largest ecosystem on Earth and supports a vast number of diverse organisms which are adapted to living in the salty environment.

The main abiotic attribute of the marine ecosystem is the presence of saltwater, with salinity levels typically around 3.5%. The salinity of the marine ecosystem affects the osmoregulation of marine organisms. The light availability determines zones like photic (sunlit), dysphotic (twilight), and aphotic (dark) which influence the types of organisms living in each zone. The temperature varies with depth, latitude, and ocean currents but stays relatively stable throughout the year. Ocean currents distribute heat globally, moderating climates on land and in the sea. Marine waters have regional variation; tropical waters are warm and biodiverse, while polar waters are cold and nutrient-rich. Water pressure increases with depth, influencing species distribution and adaptations. Nutrient levels are crucial for marine productivity. An overabundance of nutrients can be just as catastrophic as the absence of essential nutrients by creating algal blooms which deplete marine environments of oxygen resulting in dead zones where nothing can live.

Producers found in the marine ecosystem are algae, phytoplankton, kelp, and seagrasses. A large group of animals in marine waters are invertebrates; jellyfish, mollusks, starfish, sea urchins, crabs, clams, mussels, oysters, squid, and octopus. The vertebrates include bony and cartilage fish such as sting rays, sharks along with sea mammals like seals, walruses, dolphins, and whales. The aquatic biome is threatened by pollution, overfishing, climate change, habitat destruction, and invasive species. Lionfish and Asian kelp are disrupting multiple marine habitats by outcompeting other organisms and spreading their geographic location at record speeds.

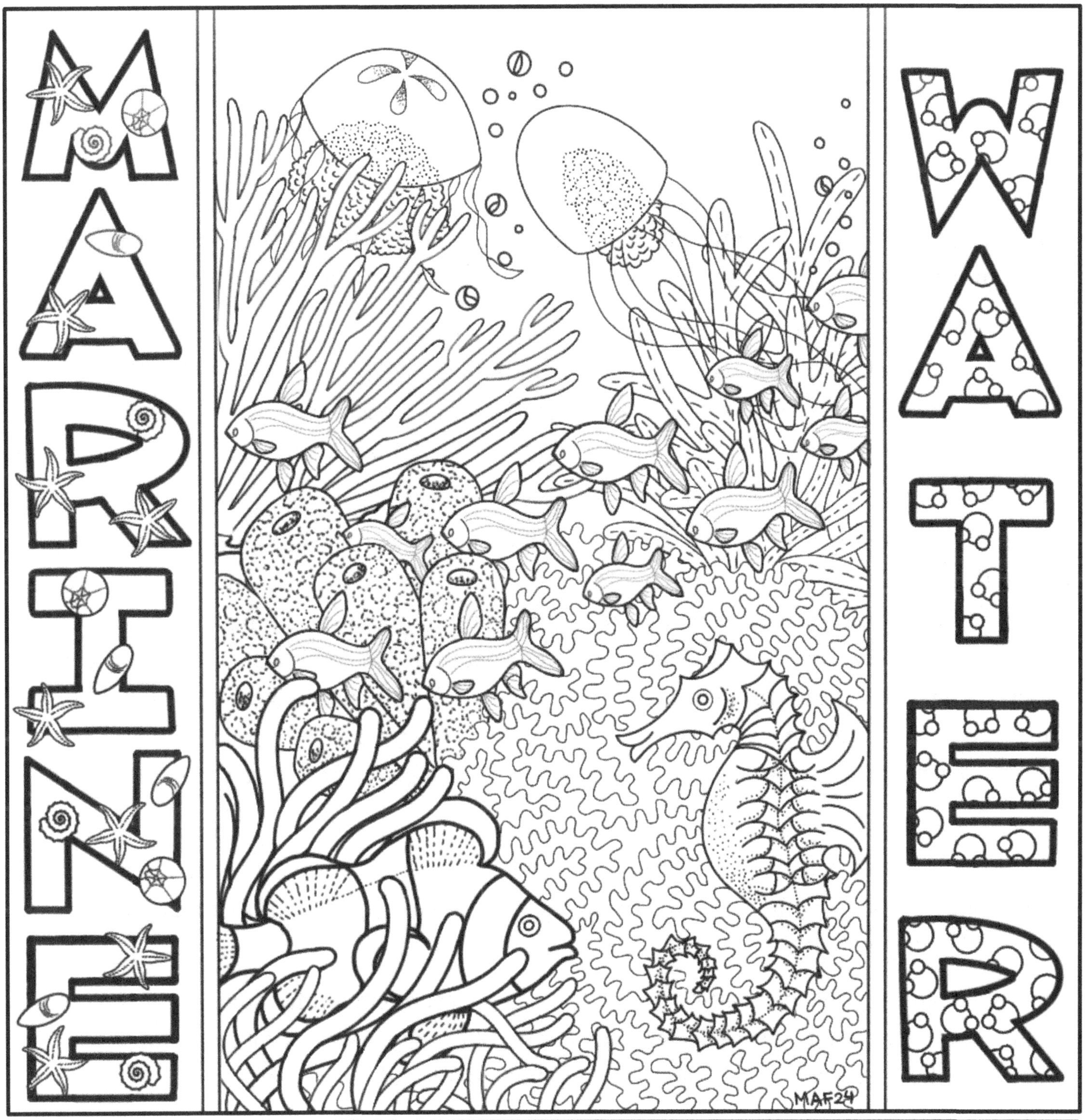

List at least three marine ecosystems.

What are the three light zones of the marine ecosystem called and describe based the light availability for each.

What are three major abiotic factors of the marine water ecosystem?

List 2 invasive species that live in marine water.

Coral Reef: A diverse marine ecosystem formed by colonies of coral polyps that secrete calcium carbonate, creating a hard, protective skeleton, the reefs foundation. Coral belong to the animal kingdom. The reefs are highly productive and serve as habitats for a vast array of marine species.

Abiotic elements of coral reefs are temperature, light availability, salinity, water clarity, and substrate. Coral reefs require warm waters between 23°C and 29°C (73.4-84.2 °F) ideally and are nutrient poor. High light and water clarity is essential for photosynthetic algae (zooxanthellae) within coral tissues. Coral reefs need stable salinity levels of around 3.5%. Coral reef require solid substrate for coral larvae to attach on and grow.

Major biotic elements of coral reefs are producers called Zooxanthellae (symbiotic algae which live on the coral framework) perform photosynthesis, supporting the coral and the ecosystem. Zooxanthellae are sensitive to temperature. Rising water temperature from global warming causes coral to expel the zooxanthellae resulting its death and in corals losing their colors to become "bleached". Many fish (e.g., parrotfish, clownfish), eels, invertebrates (e.g., starfish, crustaceans), and predators such as sharks and seabirds have made the coral reef their home. Decomposers such as bacteria and fungi are vital in recycling nutrients in coral reefs.

Coral reefs provide food, protection, and economic opportunities for coastal nations and islands. Reefs protect the coastlines from storms, erosion, tsunamis, and waves by being a natural barrier to reduce water impact. Multiple antiviral, anticancer, and anti-inflammatory medicines are harvested from coral reefs.

Coral reefs face multiple ecological threats. The rise in ocean temperature and sustained periods of time (marine heat waves) are decreasing biodiversity. In areas with coral bleaching, biodiversity has declined due to food sources being diminished. Ocean acidification is the increased levels of carbon dioxide creating acidic marine waters which harms coral reefs. Coral reefs are also adversely affected by pollution, habitat destruction from ocean floor dredging, boat damage from overfishing, chemicals in sunscreen, and severe weather events which physically damage the reef.

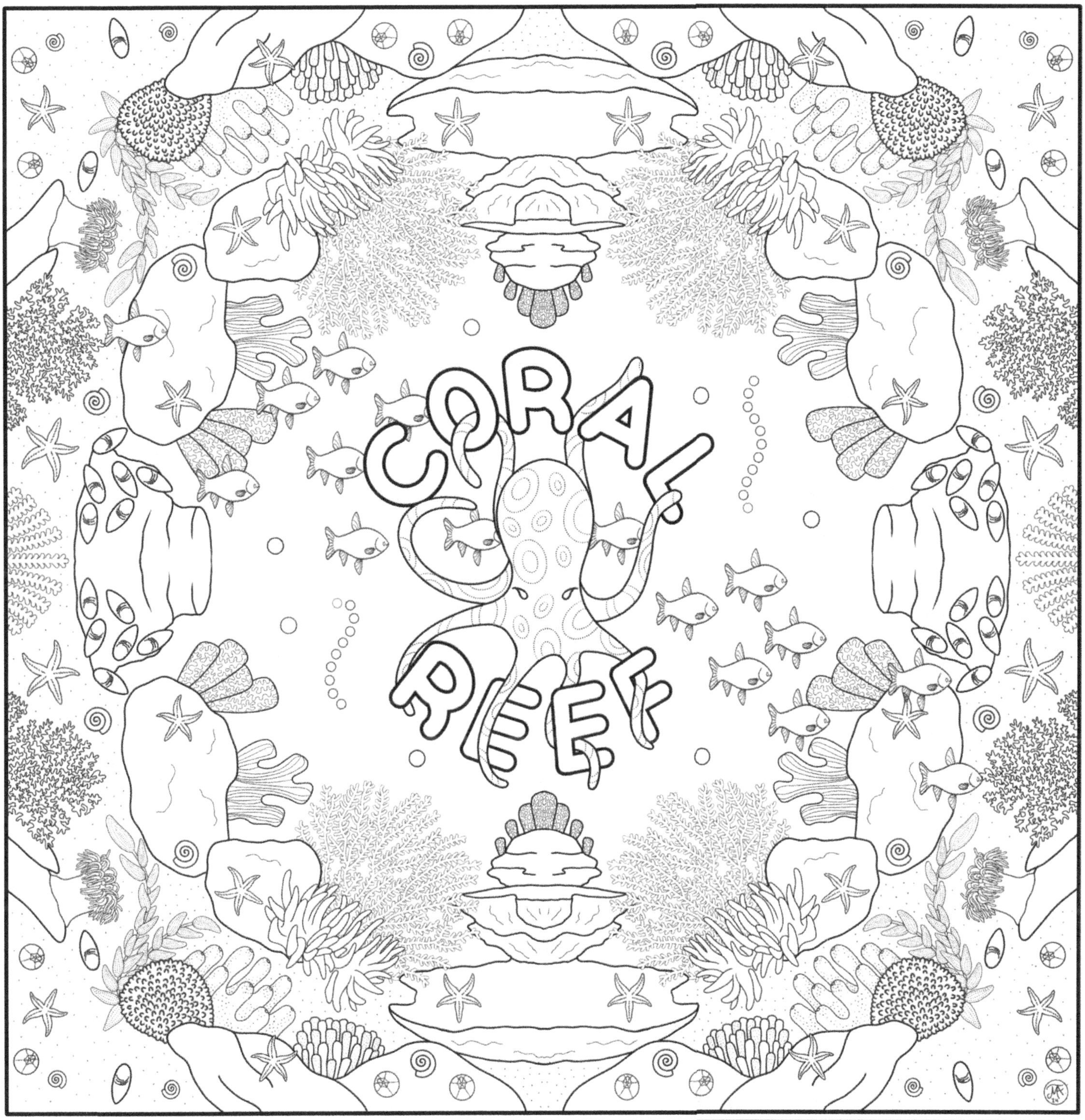

What are coral reefs made of, and how are they formed?	Why are Zooxanthellae important in coral reefs?
Which abiotic elements do coral reefs require to maintain health and biodiversity?	What is coral bleaching, and what causes it?

Kelp Forest: A dense cold aquatic ecosystem dominated by large, fast-growing seaweeds called kelp. Kelp forests occur in nutrient-rich, shallow cold coastal waters and provide critical habitats for a variety of marine species. Giant Kelp grows up to 30 meters (98 feet) tall creating a majestic underwater canopy which is productive while being an important carbon sink. Carbon sinks are areas where large amounts of carbon dioxide are absorbed to create glucose and release oxygen (photosynthesis).

Kelp forests rely on abiotic factors such as water temperature, light, nutrients, wave action and substrate. Water temperature needs to be between 5°C to 20°C (41-68°F). High light availability is important for kelp photosynthesis. Kelp require abundant nutrients such as nitrogen and phosphorous for growth. Giant kelp can grow up to 60 cm (24 inches) a day. Wave action ensures kelp stay upright and the water movement enhances their nutrient absorption. A rocky seafloor substrate creates the stable surface for kelp to anchor into. Kelp forests grow well in rocky reefs because they have long roots which attach to submerged rocks. This holdfast system allows giant kelp to withstand strong ocean currents.

Biotic elements found in kelp forests besides the giant kelp are sea urchins, snails, sea otters, rockfish, and orange garibaldi. The kelp forest is a spawning ground for many marine animals. Many animals take shelter within the kelp forest, while sea otters wrap themselves in kelp to keep from drifting away while they sleep.

Kelp forests face several threats, including climate change, which leads to rising water temperatures. Pollution from plastics, chemicals, and habitat destruction caused by coastal development also endangers these ecosystems. Overgrazing by purple sea urchins, exacerbated by a lack of natural predators, can result in barren rocky sea floors. Additionally, invasive species contribute to further damage to the kelp forest habitat. Hornweed and Asian kelp are prolific brown alga which outcompete giant kelp in their natural habitat. Controlling purple sea urchin populations, replanting kelp, and designating marine protected areas can help remediate kelp forest destruction.

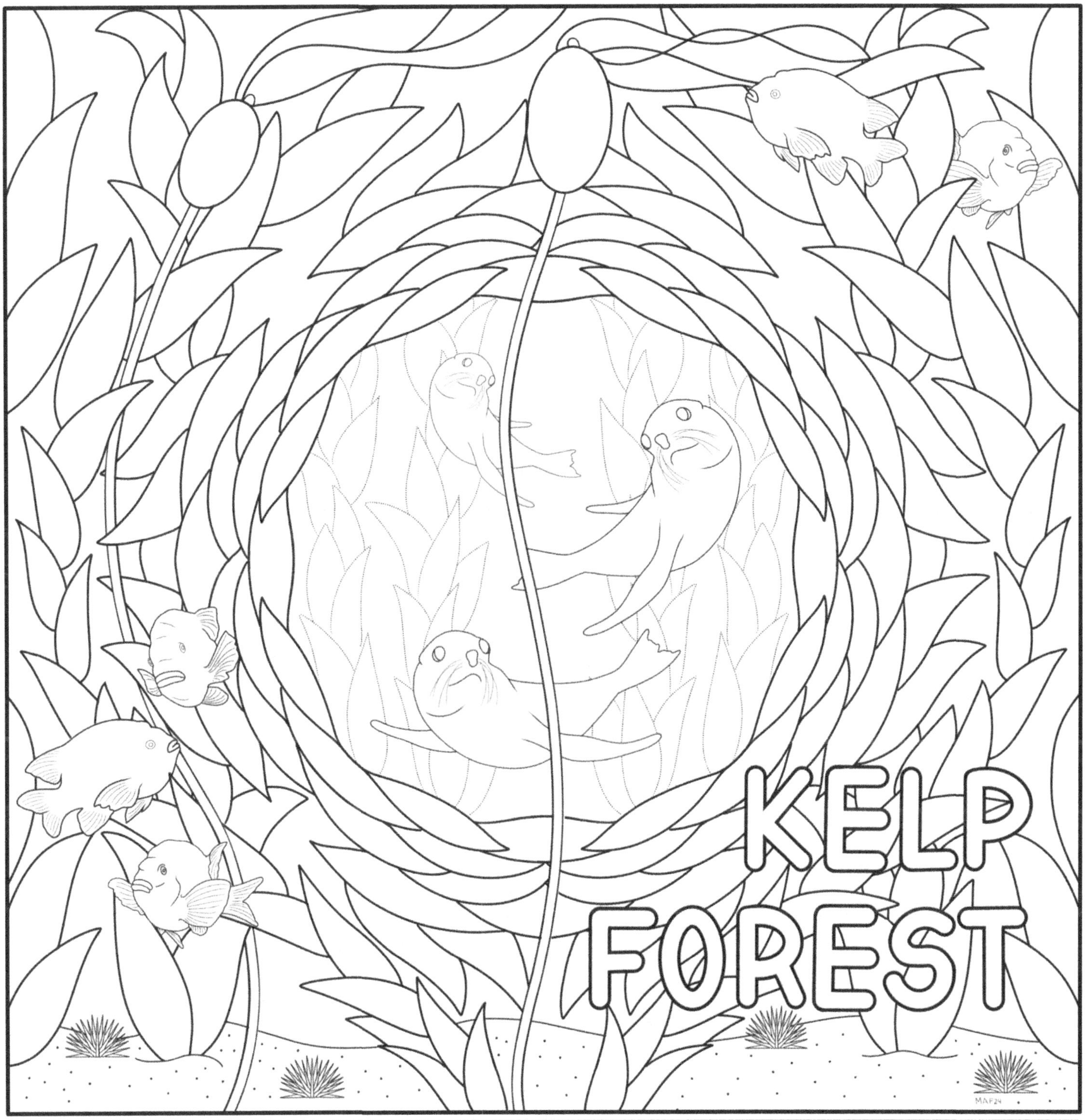

What types of water conditions are ideal for kelp forests? How is giant kelp adapted to live in the rocky reef?	Why are kelp forests carbon sinks? What are three invasive species negatively affecting the kelp forest and how are they causing damage?

Hydrothermal Vents: community, elusive community of organisms located on the dark ocean floor, centered around hydrothermal vents (geothermal fissures) where mineral-rich, superheated water is released from the Earth's crust. These communities are distinct because they thrive without sunlight, relying on chemical energy (chemosynthesis) instead of photosynthesis to sustain life. Each hydrothermal vent has a different assortment of species depending on which ocean they are located in.

The hydrothermal vents are an extreme environment with high pressure, hot temperatures, and toxic chemicals that only specialized organisms can survive in. The marine water is filled with high concentrations of hydrogen sulfide, methane, and other minerals. These inorganic molecules are used by bacteria in a process called chemosynthesis to produce glucose. Water temperatures range from near freezing in surrounding areas to over 400°C (750°F) at the vent openings. Because the hydrothermal vents are deep in the ocean the pressure from the water is immense. No sunlight reaches this unique community, and they are typically found along the tectonic plate boundaries and mid-ocean ridges. Although there is no sunlight in the area, scientists have found that dead plankton and carcasses do sink to these depths and also provide food for the tube worms, crabs, and bacteria. This means that the sunlight does still provide energy to this marine bottom community.

Organisms who live in the hydrothermal vent communities are adapted to high pressure and temperatures. Chemosynthetic bacteria and archaebacteria use the vent molecules to produce energy. Tubeworms have a symbiotic relationship with the chemosynthetic bacteria living in their tissue to help provide energy. There are also crabs, giant clams, and shrimp living in the community. Zoarcid fish and octopi also occupy hydrothermal waters as predators.

Ecological threats to the hydrothermal vents include deep-sea mining and deep-sea fishing. Ocean acidification and warming from climate change impacts vent chemistry and organism survival. Hydrothermal larvae rely on specific ocean currents for dispersal and with changes in temperature, ocean currents are affected. Microplastics and other pollutants are increasingly being found in deep-sea environments. Excessive scientific and tourist exploration damages the fragile vent structures.

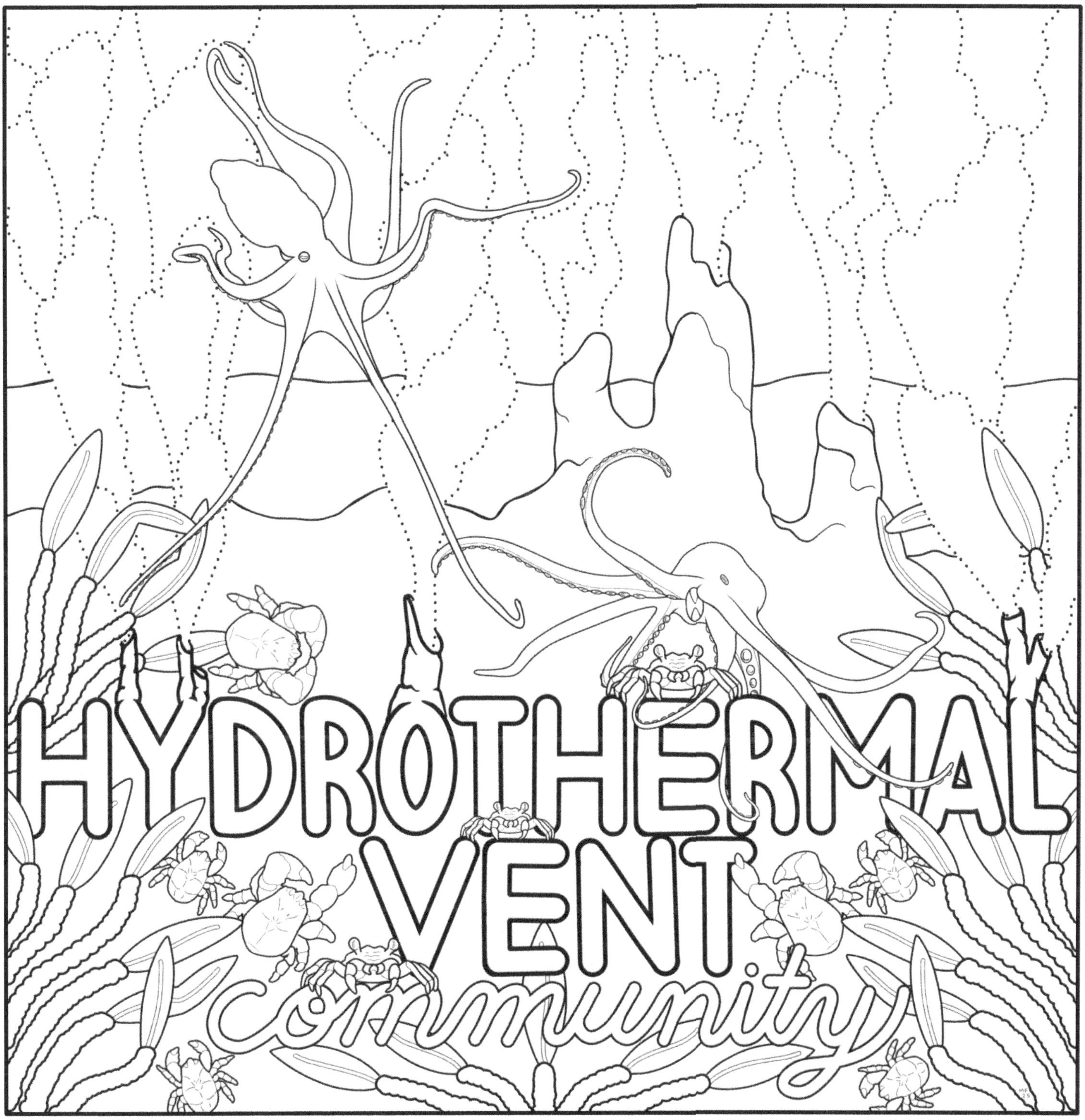

What are some unique characteristics of hydrothermal vent communities?

What is chemosynthesis?

Which organisms live near the deep-sea hydrothermal vents?

What kinds of ecological threats damage hydrothermal vents?

Estuary: An estuary is a coastal area where freshwater from rivers and streams meets and mixes with saltwater from the ocean to create a unique brackish environment. These areas are dynamic, nutrient-rich environments which are highly productive and diverse. The salinity levels fluctuate due to tidal movements and freshwater runoff.

Abiotic elements of an estuary are salinity, water temperature, nutrient levels, tidal action, and sediment. Salinity varies from freshwater to full seawater, depending on tides and freshwater input. Estuary water temperatures vary with seasons and geographic locations. Estuaries are found in tropical and temperate regions. Occasionally estuaries experience storms. Estuaries are rich in nutrients due to the mixing of ocean water and riverine input. Tides influence estuary water levels, salinity, and circulation. Fine particles carried by rivers and tides create muddy or sandy sediment bottoms.

Estuaries are full of salt-tolerant plants like mangroves, salt marsh grasses, and seagrasses, as well as phytoplankton. Estuary producers filter pollutants from freshwater before it reaches the ocean which improves water quality. Mangroves have ultra-filtration roots that remove salt from the water and specialized leaves which secrete salt. Mangroves have lateral and aerial roots to adapt to the dynamic estuary environment. Estuaries serve as protection for land against storm surges, erosion, and flooding from hurricanes. Estuaries are full of fish (e.g., salmon, flounder), saltwater crocodiles, snakes, shellfish (e.g., oysters, clams), and various invertebrates. Most animals change their behaviors and where they live depending on the salinity concentrations which varies seasonally and in some cases daily. Bacteria and fungi break down organic matter. Many migratory species use estuaries for breeding, feeding grounds, and stop over destinations.

Threats to the health of an estuary include pollution, sewage, and industrial waste which lead to degradation of water quality. Excess run-off fertilizer from agriculture results in an overgrowth of algae (algal bloom) which deplete oxygen from the water making it inhabitable. Habitat destruction from coastal development, dredging, and freshwater reclamation disrupt the ecosystem. Overfishing of fish and shellfish populations depletes the food chains. Climate change and extreme storms result in sea levels rising with salinity disturbances.

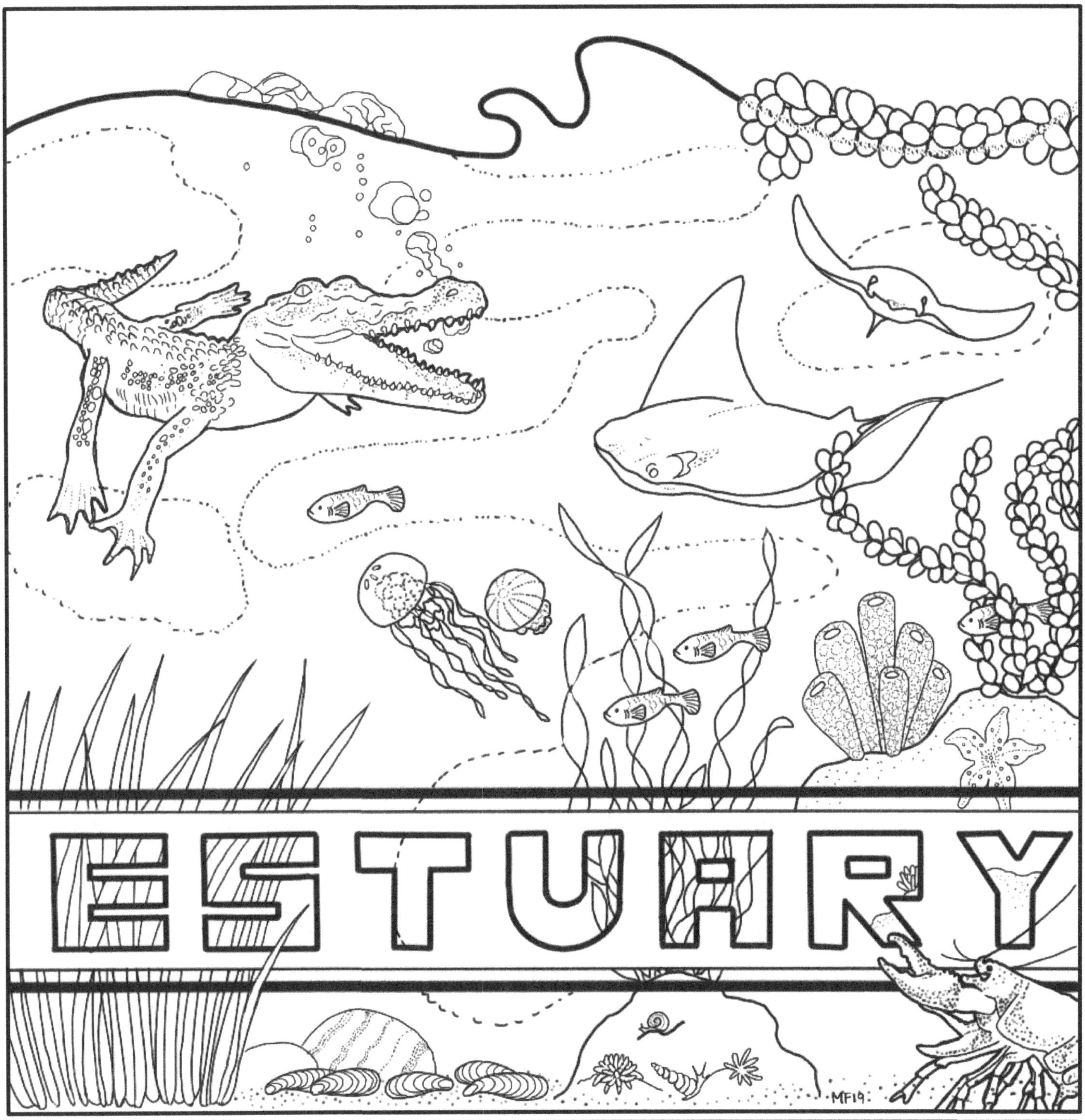

Describe estuaries.

How do tides affect estuaries?

Why are estuaries a constantly changing environment?

What kinds of adaptations do estuary plants have?

Freshwater Ecosystem: A freshwater ecosystem is a type of aquatic ecosystem that consists of freshwater bodies such as rivers, lakes, ponds, streams, and wetlands. Freshwater bodies of water have less salinity than marine. These ecosystems support a wide range of plant and animal life. Freshwater environments can be found in cold to tropical and even desert climates. The climate can influence the temperature, water levels, and seasonal variations in flow. For instance, rivers in colder regions might freeze in the winter, while lakes and ponds in warmer climates may support year-round biodiversity.

Freshwater abiotic elements include water, temperature, nutrients, light, and oxygen levels. Freshwater systems have salinity levels of less than 1%. Temperature in freshwater environments fluctuates more than in stable marine environments and are more influenced by the weather due to being smaller bodies of water and often surrounded by land. Nutrients in freshwater systems varies, levels are immediately affected by human activities such as farming and industry. Light availability is important for photosynthesis and is affected by water depth and clarity. Dissolved oxygen is critical for most freshwater organisms, low levels leads to organism die-off.

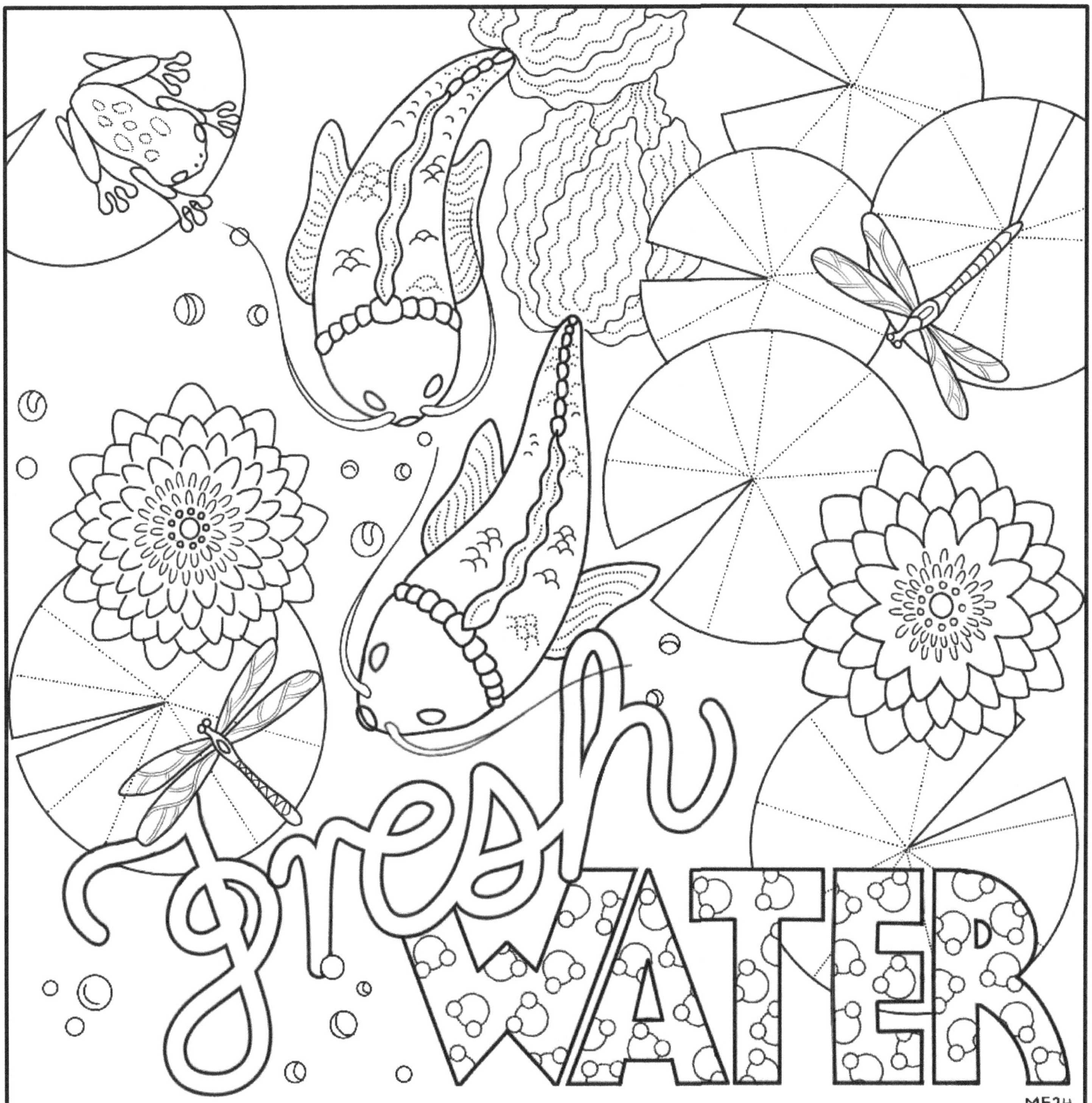

What is the salinity level of freshwater ecosystems?

List four bodies of water which are usually freshwater.

Why do freshwater temperatures fluctuate more than marine environments?

What happens when freshwater oxygen levels get too low?

River: A river is a natural flowing dynamic watercourse, usually freshwater, that moves toward an ocean, sea, lake, or another river. It is an essential component of the water cycle and plays a crucial role in shaping landscapes, providing habitat, and supporting biodiversity. A river ecosystem's flowing water moves continuously in one direction. The kinetic movement influences oxygen levels, sediment transport, and the distribution of organisms within the ecosystem.

Abiotic elements of river ecosystems are water, oxygen, temperature, sediment, minerals, pH level, and sunlight. Water is the primary medium of the river ecosystem. The temperature varies with location, season, and depth. A single river can travel through multiple ecosystems and therefore have a wide range of temperature fluctuations. The oxygen levels are usually high due to the movement of water. Sediments and minerals within the river affect water clarity and nutrient availability. Seasons affect the ratio of sediment to water concentrations. The pH levels determine the chemical balance of the river and can be influenced by the sediment and nutrients. Sunlight influences photosynthesis in aquatic plants and algae. Sunlight has UV rays which help kill pathogens in the water while also helping to break down pollutants.

The river ecosystem has a range of producers from microscopic to large trees. Algae, aquatic plants, phytoplankton grasses, water lilies, cattails, reed grasses, aquatic adapted trees provide oxygen and food. Rivers have a large array of animals from fish, amphibians, insects, and even aquatic mammals and reptiles. Fish such as trout, salmon, and catfish can be found in rivers. Amphibians such as frogs, toads, salamanders, and newts spend their entire life dependent on the river ecosystem. Mammals such as otters and beavers live in the river along with crocodiles, alligators, turtles, and snakes. Bacteria, zooplankton, and fungi break down organic material to recycle nutrients.

Rivers exist in nearly every climate barring the extremely cold and extremely hot and dry climates. Rivers can not flow in extremely cold climates, occasionally polar climates may have some temporary water runoff but no permanent rivers. Extremely hot dry deserts can not sustain rivers, they dry up too fast. The Nile river in the Sahara desert is an exception due to its size and volume of water. The Everglades in Florida is a slow wide and shallow river system. It flows at the rate of 0.6 meters (2 feet) per minute 30–90 meters (100-300 feet) per day. The flat terrain and concentration of grasses keeps the Everglades from flowing faster. The Everglades has a depth of a few centimeters to about a meter.

Ecological threats to rivers are pollution from industrial waste, agricultural runoff, human irresponsibility, and plastics. Dams and water diversions disrupt natural flow, sediment transport, and fish migration. Deforestation increases erosion and sedimentation, altering water clarity and flow. Climate change alters precipitation patterns, leading to droughts or floods. The introduction of invasive species and overuse of water dissrupt the fragile ecosystem.

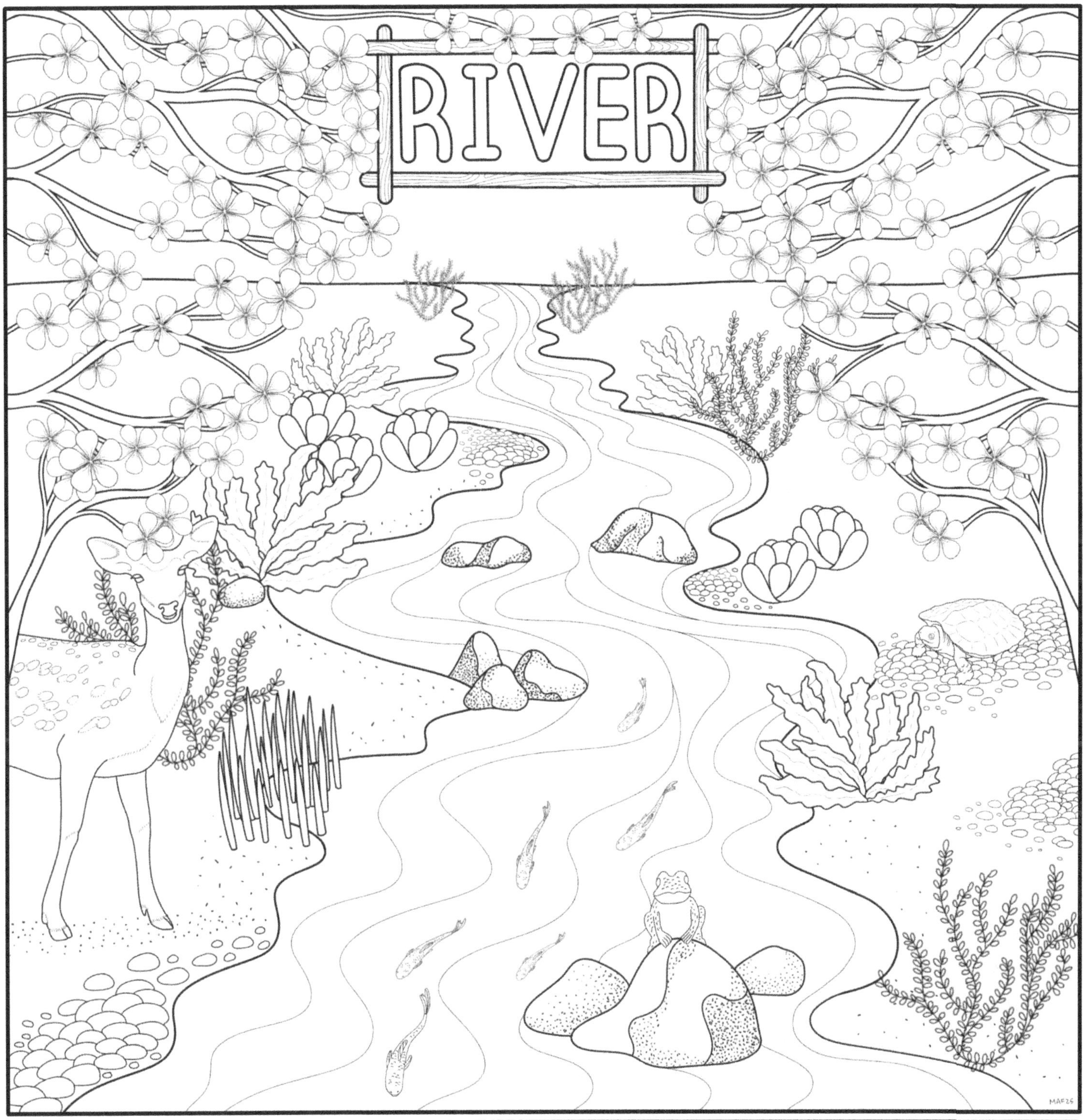

What is a river?	List three abiotic features of a river.
Why are rivers not found in extremely cold climates and extremely hit climates?	What are two ways human activities negatively affect rivers?

Lake: A lake is a large, inland body of standing water, usually freshwater, that is surrounded by land. Lakes can form through geological processes such as glaciation, tectonic activity, or volcanic activity. Some lakes form above subterranean springs which have breached the Earth's crust. They vary in size and depth and are critical habitats for diverse plant and animal species.

Lake ecosystems are divided into different zones based on depth and sunlight penetration. The littoral zone is shallow, near the shore and has an abundance of plant and animal life. The limnetic zone is open water, well-lit, primary area for photosynthesis of diatoms. The profundal zone is deep dark cold water with lower oxygen levels than the higher levels. The benthic zone is the lake bottom, home to decomposers and bottom-dwelling organisms. Abiotic elements of lakes are water, sunlight, temperature, oxygen levels, nutrients and sediment. Standing water is the main medium of lake ecosystems. The sunlight affects the temperature and rate of photosynthesis in the lake. Lakes have seasonal temperature variations from thermal stratification (layering of temperatures) in summer and mixing in spring/fall. The oxygen levels determine whether life can be sustained in the lake water. Nutrients also help determine if life can be successful in lake water. Too many or too little nutrients will not sustain life, there must be a good balance. Sediment affects water clarity and habitat structure. Some lake sediment is sandy and soft, while other lakes have rocky sediment and coarse material.

Lake ecosystems support a variety of aquatic plants that are adapted to different water depths and conditions. These plants are classified into submerged, floating, and emergent plants based on where they grow within the lake. Submerged plants are entirely underwater and are rooted in the lakebed. They play a crucial role in oxygenating the water and providing habitat for fish and invertebrates. Elodea, hydrilla, and eelgrass are some examples. Floating plants float on the water's surface and have roots that dangle in the water or are free-floating. Water lilies, duckweed, water hyacinth and lotus are important floating plants which provide shade and habitats. Emergent plants are rooted in the lakebed and growing above water, their roots help to stabilize the sediment and prevent erosion. Cattails, bulrushes, pickerelweed, and arrowhead provide habitats and food for birds, insects, amphibians and fish. Algae and phytoplankton (diatoms) are microscopic or simple aquatic plants which form the base of the lake food web.. Lake plants filter and absorb pollutants, improve water quality, and support biodiversity. Lakes support a range of fish, amphibians, birds, and water dependent mammals such as beavers and otters.

Lake ecosystems face challenges from climate change, pollution, eutrophication, and disruptions in weather patterns which reduce or increase precipitation. Eutrophication is caused by excess nutrients from agriculture and sewage causing algal blooms, depleting oxygen and harming aquatic life. Invasive species and overfishing creates imbalances in lake ecosystems. To protect lakes, pollution must be reduced and eliminated, invasive species must be controlled, water use must be sustainable, the lake ecosystem must be conserved with public awareness and policy.

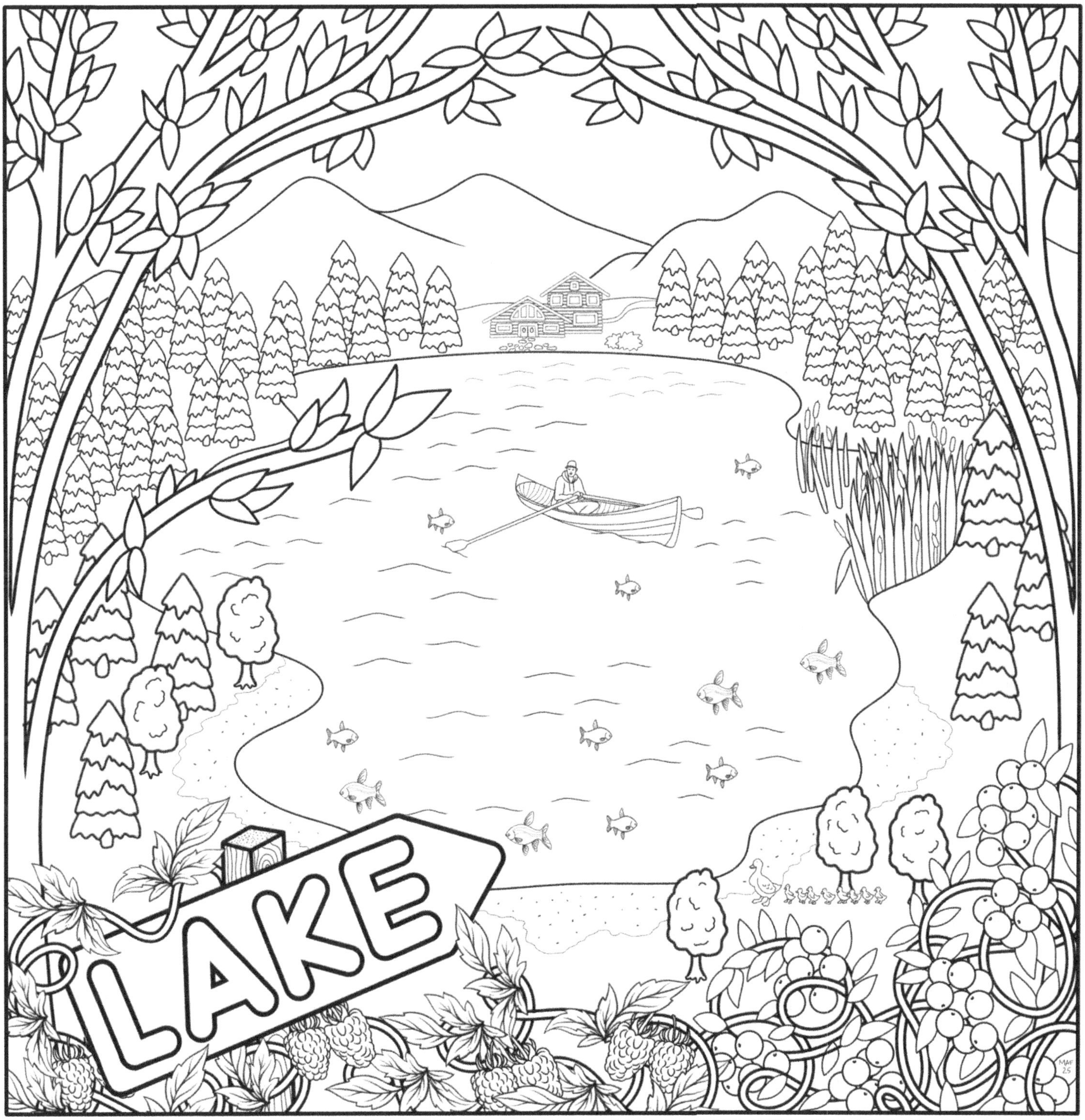

What is a lake?

What are the three different zones of a lake ecosystem?

How do aquatic plants benefit lake ecosystems?

What are two ways human activities negatively affect lakes?

Wetlands: The main characteristic of wetlands is the presence of water-saturated soil permanently or seasonally. This waterlogged environment creates unique conditions that support specially adapted plants and animals, making wetlands among the most productive ecosystems on Earth. Wetland areas include swamps, marshes, bogs, and fens, and they play a vital role in supporting biodiversity, regulating water flow, and filtering pollutants. Wetlands also store excess water which is vital during droughts.

Abiotic elements in wetlands are water, soil, light, nutrients and hydrology. Water in wetlands may be fresh, brackish, or salty. The soil will be water-logged, nutrient-rich, and often oxygen-poor such as peat or clay. The availability of light depends on the density of vegetation and water clarity. Wetlands have high levels of nutrients depending on the type of wetland and season. Hydrology is the seasonal or permanent water levels which are influenced by precipitation, tides, storms, and groundwater.

Producers of wetlands are water tolerant plants such as cattails, reeds, mangroves, and algae. There are many primary consumers like muskrats, rabbits, a variety of birds and insects. Ducks, geese and other migratory birds also frequent wetlands. Freshwater carnivores such as alligators, crocodiles, and snakes are also well adapted to the wetlands, including the Burmese python which has become invasive. A large range of fish, frogs, toads, and salamanders are also at home in the wetlands. Microorganisms such as bacteria, fungi, and protists aid in decomposition for nutrient cycling.

Wetlands can exist in nearly any climate. The climate influences the wetland's water levels, temperature, and the species it supports. Tropical wetlands have warm temperatures and high rainfall year-round. Temperate wetlands have seasonal variation in temperature and water levels. Arctic wetlands have short growing seasons and frozen conditions for most of the year.

Ecological threats to wetlands are land development. Many natural wetlands have been drained for suburbanization only to be detrimentally flooded from future storms. Invasive species such as alligator gar, Burmese pythons, and carp outcompete native species and upsetting the natural food webs.

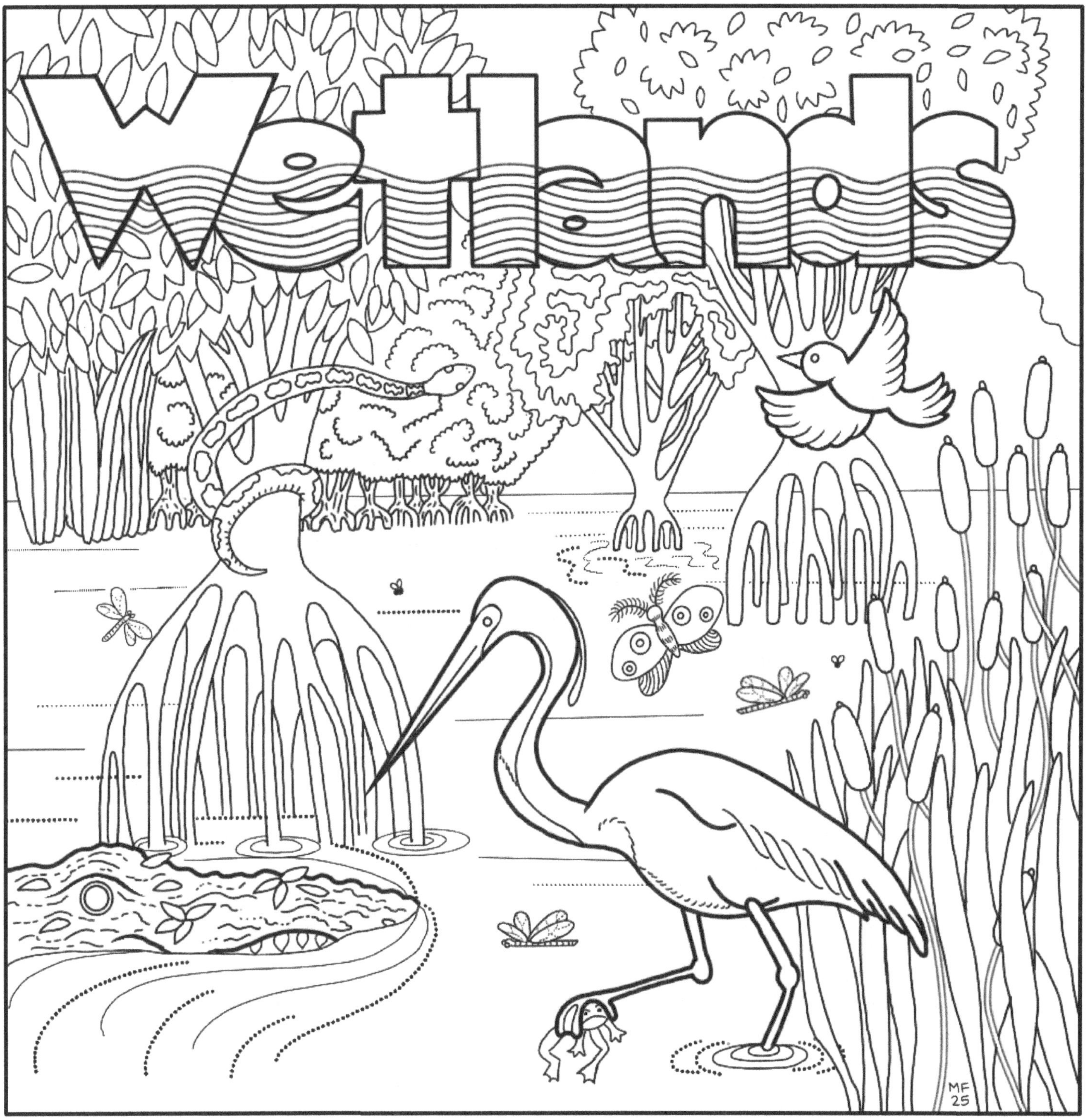

Describe the wetlands.

How do wetlands protect land and water?

What are the ecological threats to wetlands?

What kinds of invasive species can be found in wetlands?

Grassland Biome: Ecosystems dominated by grasses, wildflowers, and other non-woody plants, with few trees or shrubs. They are found on every continent except Antarctica and include savannas, prairies, steppes, and pampas. Vegetation on grasslands are limited by factors like low rainfall, frequent fires, and grazing by herbivores. Grasslands have high levels of pollination and are buffers against extreme weather events such as droughts and flooding.

Abiotic elements in grasslands are soil, water, sunlight, temperature, and wind. Soil in grasslands are often nutrient-rich from the decomposition of grasses and the fertilizing manure from herbivores. There are some grasslands with poor soil depending on the region. Seasonal precipitation ranges from 25-76 cm (10 to 30 inches) per year. Droughts are common and many grasslands have fire seasons. Grasslands receive ample sunlight which is vital for grass growth. Temperatures vary from very warm to cold. Tropical grasslands are warm throughout the year while temperate grasslands have warm summers and cold winters. Winds help shape the landscapes by dispersing seeds.

Biotic elements of grasslands include herbivores of all sizes. The largest herbivores on Earth, elephants & giraffes, live in grasslands along with bison, antelope, zebras, and kangaroos. Carnivores such as lions, cheetahs, coyotes, and wolves hunt in the tall grasses. Grassland decomposers are fungi, bacteria, and insects like beetles that recycle nutrients into the soil. Grasslands help mitigate climate change because they are carbon storing landscapes with large amounts of carbon dioxide within the grasses. Grasslands also aid in water regulation because the roots help trap rainwater and prevent erosion.

Grasslands are threatened by agriculture, urbanization, overgrazing, climate change, invasive species, and deforestation. The conversion of grasslands into farmland or pastures leads to habitat loss and soil degradation in part due to nonnative crops being planted. Native crops have the adaptations such as deep spreading roots and regenerative properties after fires. Nonnative plants have shallow root systems and are permanently wiped out by fires leading to soil degradation, erosion, and inability to store large quantities of water. The urbanization of expanding cities encroach on grassland areas. Overgrazing by livestock reduces vegetation and leads to soil erosion. Deforestation affects water cycles and soil stability in grasslands.

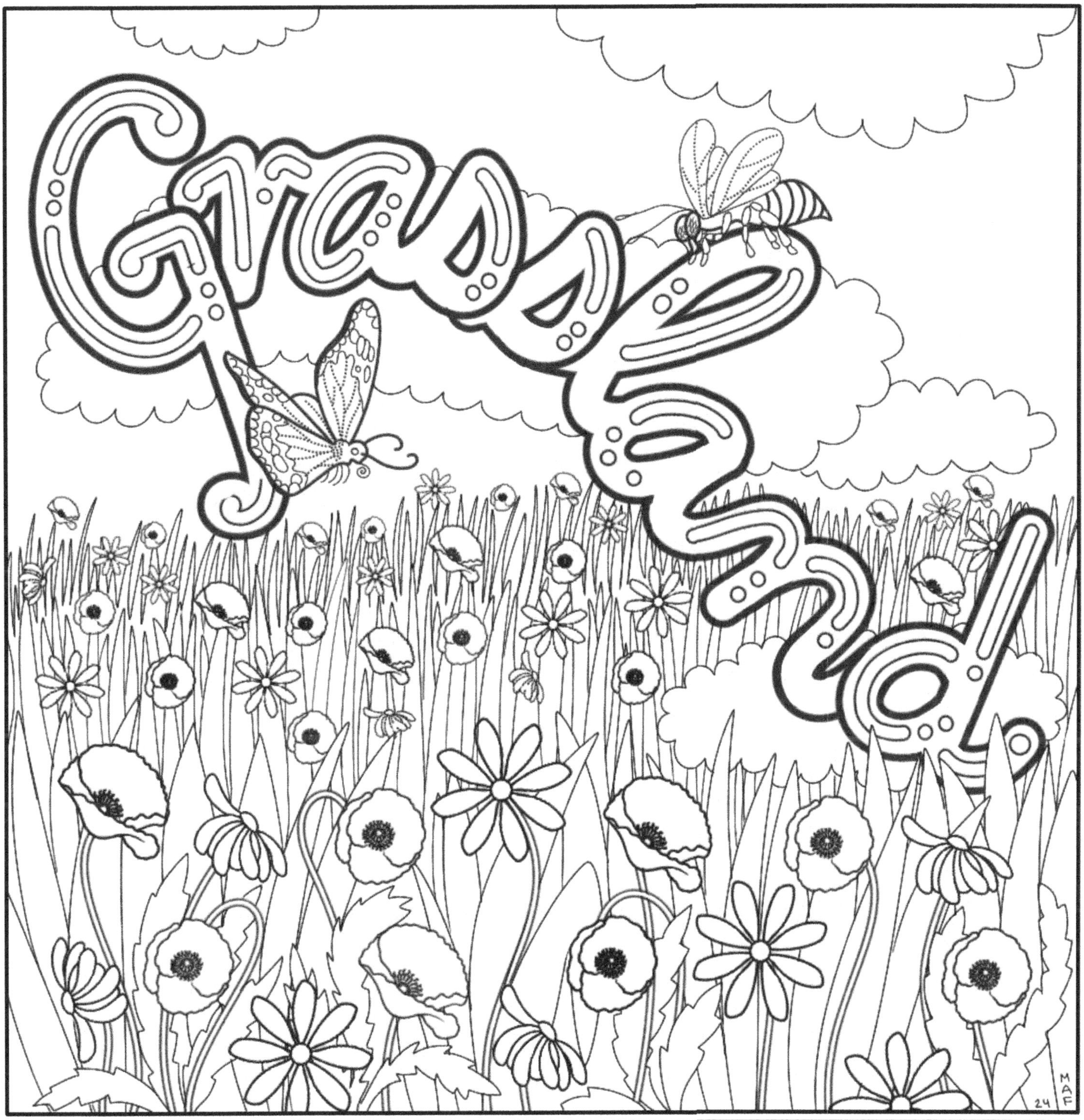

What are the main characteristics of grasslands?

What are three abiotic factors of grasslands?

How do grasslands help mitigate climate change?

What contributes to the soil erosion in the grasslands?

Temperate Grassland: Ecosystem dominated by grasses and herbaceous plants, with few trees due to limited rainfall. These ecosystems are found in regions with moderate climates and are known for their fertile soils, making them prime areas for agriculture. Temperate grasslands are shaped by moderate rainfall, frequent fires, and grazing by herbivores. Temperate grasslands are located in regions with moderate climates, such as the Great Plains in North America, the Pampas in South America.

Temperate grassland abiotic factors include soil, water, light, wind, and temperature. Temperate soil is nutrient-rich and deep from centuries of herbivores grazing and decomposing of bacteria, fungi and insects. Seasonal rainfall ranges from 25-76 cm (10 to 30 inches) per year. Temperate grasslands have hot summers and cold winters. There is plenty of light for growing the grasses and wind for seed dispersal.

Biotic elements include a variety of grasses, wheatgrass, wildflowers and few trees. Bison, antelope, prairie dogs, and grasshoppers are prevalent. Predator species such as coyotes, wolves, and birds of prey such as hawks also live in temperate grasslands. Animals are various shades of brown to camouflage. Some of the animals burrow to escape temperature swings and predators. Many species migrate through this area with the season changes. The migration of large animals and excrement they leave behind, gives the temperate grasslands one of the most nutrient rich soils in the world. The grasses have deep root systems (3-4 meters deep, 9-12 feet) which allow them to access water during droughts and prevent damage from strong winds and large herbivore grazing. Grasslands have high rates of photosynthesis due to the fast growth of various grasses and wildflowers which results in a high rate of carbon absorption and conversion into glucose.

Conversion of grasslands into monocrop industrial farms depletes soil nutrients and reduces native biodiversity. Excessive grazing by livestock can lead to soil erosion and degradation. Expanding cities and infrastructure encroach on grasslands destroying habitat for grasses and animals. Climate change alters rainfall patterns and increases temperatures which affects grass growth and biodiversity. Fire suppression by humans disrupts natural cycles, allowing shrubs and trees to encroach and take land from grasses.

TEMPERATE Grassland

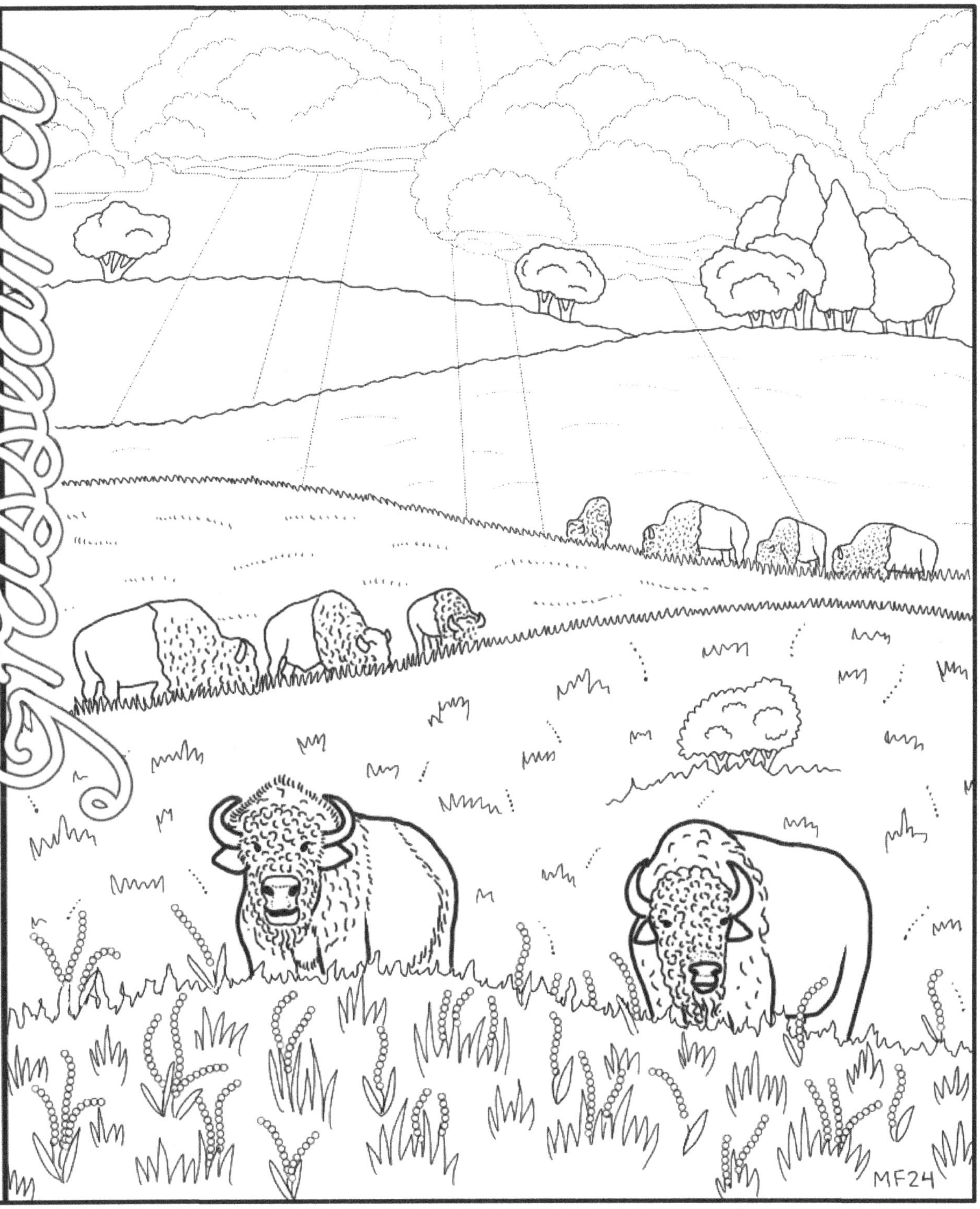

Where are temperate grasslands typically located?

Why is the temperate grassland soil so rich?\

What are some examples of animals and plants that live in temperate grasslands?

How can temperate grasslands be protected and restored?

Savanna: Tropical or subtropical grassland ecosystems characterized by a mix of grasses, shrubs, and scattered trees, often found in regions with seasonal rainfall. They are commonly located between tropical rainforests and deserts while prevalent in parts of Africa, South America, Australia, and India. The landscape is being shaped by seasonal rainfall, fire, and grazing. The trees are usually widely spaced and umbrella-shaped allowing ample sunlight for grasses to thrive.

Savannas typically have nutrient-poor soil but can support a variety of plants. The seasonal rainfall releases 50 to 152 cm (20 to 60 inches) of water annually then followed by a long dry season. Year-round the savanna is warm with temperatures averaging between 68°F and 86°F (20°C to 30°C). Savannas have ample amounts of sunlight for the grow season and frequent fires which prevent tree overgrowth while promoting the growth of fire-resistant grasses.

Savannas are home to at least 75 species of grasses such as common finger grass, lovegrass, red oat, and Bermuda grass, along with scattered trees such as acacia and baobab. Plants of the savanna have extensive deep root systems that can reach groundwater to survive drought periods. They limit flowering to specific seasons to conserve energy. Some plants develop specialized structures such as bulbs, stems, and tissues to store water. Savanna plants have fire tolerant bark and deep roots that are protected from fires so the grasses can quickly regrow following fires while withstanding herd grazing.

Savannas are home to large mammals such as elephants, zebras, giraffes, and antelope. African elephants are adapted to use their large ears to keep cool. Giraffes use their long necks to reach the leaves of the acacias. Giraffes have long black slime coated tongues, the color protects their tongues from the sun while the slime allows them to eat thorny plant material. Zebras and antelopes migrate to eat and obtain water during the drought season. Many small savanna animal species have learned to hibernate, burrow or become dormant to survive the dry spells. Many have strong front limbs with claws to help them burrow underground. Savannas have multiple carnivores such as lions, cheetahs, hyenas, and wild dogs, which prey on herbivores. Savanna decomposers; fungi, bacteria, and insects break down organic matter and recycle nutrients back into the soil. Animal adaptations of the savanna also include animal coloring and patterns that help them camouflage into their surrounding.

Climate change threatens the savanna with altered rainfall patterns and prolonged droughts. Overgrazing of livestock reduces vegetation, soil fertility, and leads to desertification. Implementing rotational grazing, avoiding overgrazing, and using sustainable farming techniques can help protect soil and vegetation. Deforestation around savannas is also disruptive by clearing of land for agriculture, logging, or urban development which reduces forest tree cover and disrupts the ecosystem's weather patterns. Fire management using controlled burns and better management of natural fires can help maintain the health of the ecosystem while preventing uncontrolled wildfires.

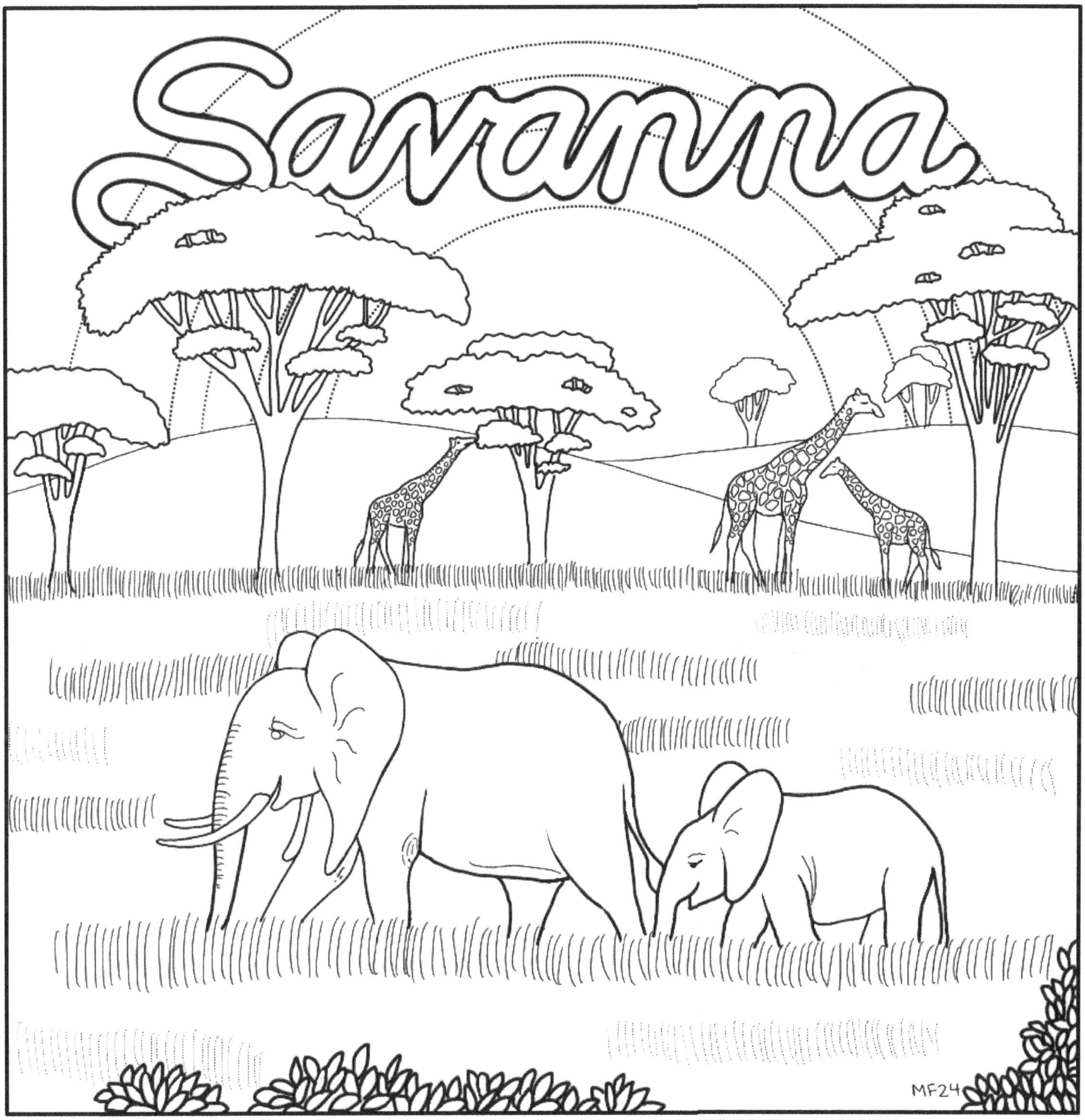

Describe the savanna ecosystem.	What kinds of adaptations do savanna animals have to survive the drought period?
What kinds of adaptations do savanna plants have to survive the drought period?	How is climate change affecting savannas?

Chaparral: Dense, shrub-dominated ecosystems characterized by hot, dry summers and mild, wet winters. Chaparrals are commonly found in regions with a Mediterranean climate such as Mediterranean countries, Western Coast of California, Southwest Australia, Chili, and South Africa. Chaparrals have low-growing shrubs, grasses, seasonal wildflowers, and small trees, with vegetation highly adapted to survive dry conditions, periodic fires, and nutrient-poor soils.

Abiotic elements of a chaparral are soil, water, temperature, sunlight, and fire. The soil is thin, rocky, and nutrient-poor but well-drained. Chaparrals have limited water availability during the long, dry summers. But can receive 25-102 cm (10-40 inches) of rain in the winter months. Summers can be hot and dry with temperatures soaring to 30-40°C (86-100°F) Sunlight is abundant year-round, promoting photosynthesis in plants. Natural and periodic fires play a significant role in maintaining the chaparral ecosystem. Too much fire suppression, preventing natural fires, can lead to an overgrowth of vegetation, increasing the risk of catastrophic wildfires. Chaparral fires naturally occur at 30–150-year intervals, but due to climate change resulting in stronger winds and hotter temperatures the intervals has significantly decreased.

Drought-tolerant and fire-resistant plants of the chaparrals include shrubs like manzanita, chamise, scrub oak, and drought-resistant plants such as sage and yucca. Chaparral trees are mountain mahogany, poison oak, oak, Tecate cypress, and some evergreens. Multiple plant species require fire to regerminate like California` lilac and Tecate cypress. The fire will melt the resin from the seed coats. Some plants will only regrow after a fire from Most underground root systems. The chaparral plants have deep root systems, with tap roots which prevent soil erosion on steep slopes. The vegetation helps to capture and retain rainwater which contribute to the watershed and regulate stream flow. Invasive species wild mustard weed is outcompeting many chaparral plants creating instability. Herbivores such as deer, rabbits, rodents, and insects along with carnivores like coyotes, bobcats, mountain lions, and predatory birds live in the chaparral hills. There are many reptiles and amphibians that make the chaparral home, both along with small mammals, burrow in the ground when it is hot or dry. A lot of animals of the chaparral have nocturnal or diurnal habits. Most chaparral animals have camouflage coloring or contrasting patterns to scare potential predators. Decomposers such as fungi, bacteria, and insects recycle nutrients from plant and animal matter.

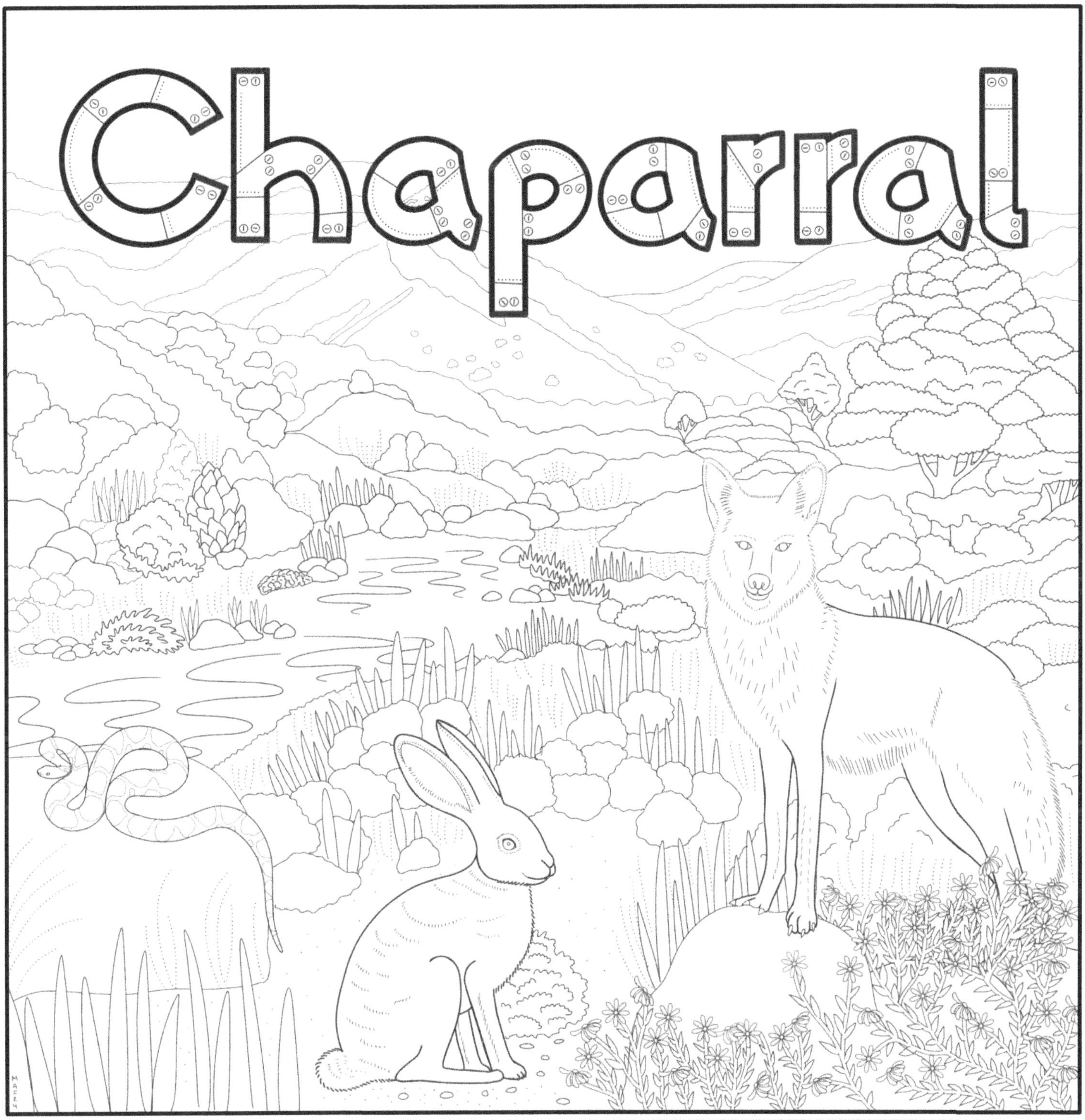

Describe the chaparral ecosystem.

What are the abiotic factors that shape chaparral ecosystems?

What fire-resistant adaptations have chaparral plants evolved?

What is contributing to the increased fires in the chaparral?

Steppe: Large, treeless plains dominated by grasses and herbs, often found in regions with semi-arid climates. Steppes are known for their vast, open landscapes and are vital ecosystems that support grazing animals and unique plant species. Steppe plants and animals are adapted to survive in dry and windy conditions with low to moderate rainfall.

Steppe soil is fertile but dry and prone to erosion and often rich in nutrients from migrating animals. Water is limited with low rainfall and occasional rivers or streams. Annual precipitation, ranges from 25-50 cm (10 to 20 in). Steppes experience seasonal temperature extremes. Summers are warm to hot (68°F to 86°F or 20°C to 30°C), while winters are cold, often with temperatures below freezing. There is ample sunlight, essential for grass growth. Strong winds are common and contribute to soil erosion and the types of plants and animals that can live there successfully.

Steppe producers are drought-resistant with deep rooted grasses like feather grass and needle grass, along with herbs and wildflowers. Animals found in steppes include grazing animals such as antelope, small horses, small deer, gazelles, bison, and small burrowing mammals like prairie dogs and wild hamsters. Carnivores like wolves, foxes, eagles, and other predators make the steppe home. Fungi, bacteria, and insects that break down organic matter and enrich the soil.

Historically the Eurasian steppe was an important route for nomadic people for trades, travel, and even warfare. Overgrazing, mining, agriculture, urban expansion, climate change, and invasive species threaten the steppe ecosystem by fragmenting the ecosystem and disrupting wildlife migration patterns. Managing livestock grazing, habitat protection, and limiting agriculture development will help to protect the steppe ecosystem. Soil conservation by implementing erosion control techniques, such as planting cover crops and creating windbreaks will help mitigate soil erosion.

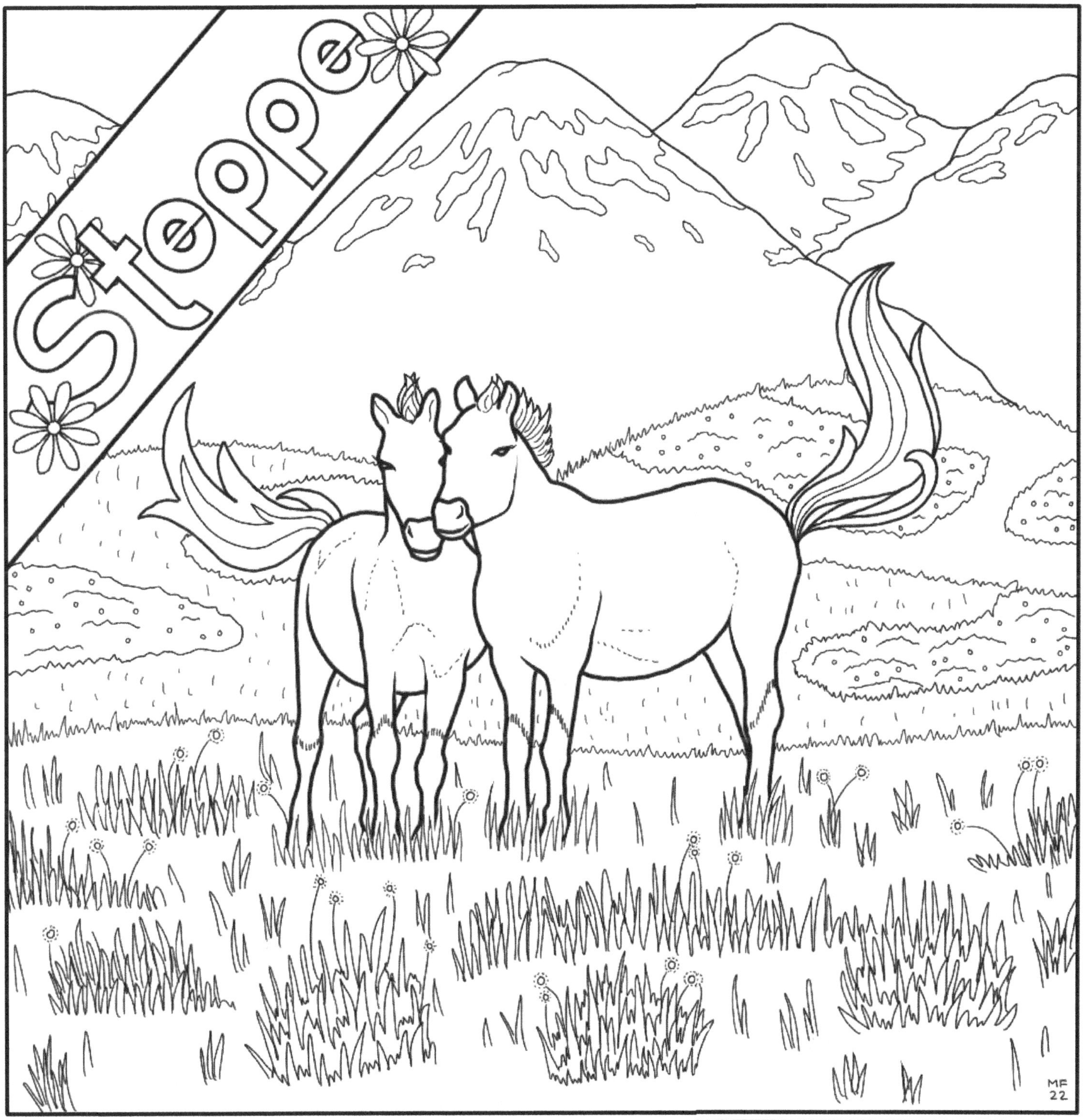

What is a steppe, and why is it important?

What types of plants and animals are commonly found in the steppe?

What are the abiotic elements of the steppe ecosystem?

Which ecological threats affect steppe ecosystems?

Forest Biome: A large terrestrial ecosystem dominated by trees and other vegetation. Forests play a crucial role in regulating the Earth's climate, supporting biodiversity, and providing resources like oxygen, food, and timber. Forests have dense tree cover and layered vegetation, ranging from towering canopy trees to shrubs, herbs, and mosses. It is highly productive and diverse, hosting numerous species of plants, animals, and microorganisms. Forests collectively cover about 31% of the Earth's land area, providing essential ecological, economic, and cultural benefits.

Forest soil ranges from nutrient-rich in temperate forests to nutrient-poor in tropical rainforests. Water availability varies by type of forest; tropical forests have abundant rainfall, while temperate forests experience seasonal variations. Forest temperatures are influenced by the forest type (tropical, temperate, boreal). Sunlight varies within the forest layers, with the canopy receiving the most sun and the ground receiving the least. Forest climate is determined by location and forest type, influencing rainfall, temperature, and seasonal patterns.

Forests have four layers of organization; emergent, canopy, understory, and forest floor. The emergent layer is where the tallest trees peak out of the forest. Birds, and insects may be found in the emergent layer, but no animals. The canopy layer is the thickest and most dense with a high number of organisms. The understory layer is composed of young saplings, shrubs, and bushes where multiple large animals interact with plants. Finally, the forest floor layer is the darkest and moist with insects, fungus, mosses, and detritus.

Forest trees can be classified as deciduous or evergreens. Deciduous trees become dormant and lose their leaves from fall to winter. Evergreens photosynthesize all year and keep their leaves. Forests also have shrubs, mosses, ferns, berries, fruit trees, and lichens. Forest consumers consist of herbivores: deer, squirrels, insects, reptiles, amphibians, and monkeys. Forest carnivores are wolves, big cats, snakes, lizards, some amphibians, and birds of prey. Forest omnivores are bears and raccoons. Decomposers which recycle nutrients are fungi, bacteria, and detritivores like earthworms.

Forests face the threat of deforestation from clearing lands for agriculture, logging, and urban development. Climate change threatens forests by altering rainfall patterns and increasing the frequency of extreme weather events. Pollution of air, water, and soil harm plant and animal health. Invasive species such as non-native plants and animals outcompete native species. Overexploitation through unsustainable harvesting of wood and other forest resources drastically harm forest ecosystems. Forest habitat fragmentation divides forests into smaller patches reduces biodiversity and forest continuity. The forest biome can be protected and restored through reforestation, afforestation, sustainable practices, conservation policies, climate action, education, awareness and community involvement.

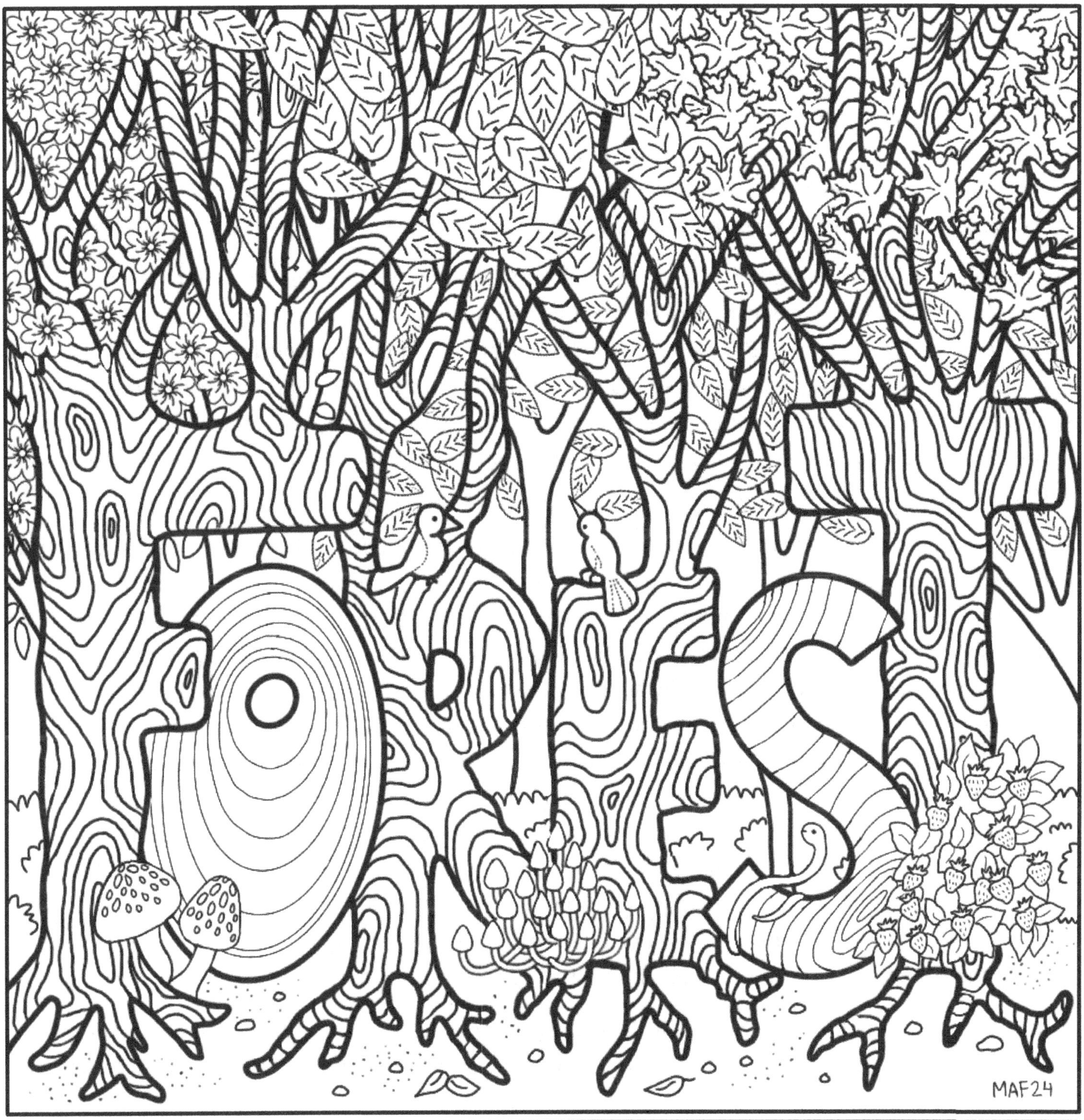

What is the dominant feature of a forest and how much land do forests cover?	What is the difference between deciduous and evergreen trees?
What are the four levels of biodiversity of a forest?	What actions can be taken to protect forests?

Tropical Rainforest: A forest ecosystem found near the equator, characterized by warm temperatures, high humidity, and abundant rainfall. Tropical forests are the most biologically diverse ecosystems on Earth, hosting an incredible variety of plants, animals, and microorganisms. Tropical forests have dense, multi-layered vegetation with towering trees forming a canopy, an understory of smaller plants, and a forest floor rich in nutrients and biodiversity. Tropical forests thrive in warm, wet conditions year-round.

Soil in tropical rainforests is often nutrient-poor and acidic, with rapid nutrient recycling. The soil is poor in nutrients because so many plants are constantly absorbing the nutrients for growth. There are high levels of rainfall, ranging from 200-1,000 cm (80 to 400 inches) annually. Tropical temperatures are consistently warm, averaging between 20°C and 34°C (68°F- 93°F). Tropical forests receive intense sunlight, although the forest floor receives limited light due to the canopy and broad leaves. Tropical humidity is extremely high, often above 80%. Tropical rainforests have minimal seasonal changes, often with a distinct wet and dry season.

Tropical rainforests have a wide range of trees such as mahogany, kapok, rubber trees, palms, and broad-leaved specimen. Many plants have broad leaves to be able to capture sufficient sunlight. Some palm trees not only have broad leaves, but they have evolved specialized roots which allow them to "walk" to sunnier spots in the rainforest. Epiphytes, plants which grow on trees, such as orchids and bromeliads are prevalent along with understory ferns. By adapting to grow on trees epiphytes received enough sunlight. Foods such as acai berries, bananas, avocados, citrus fruits, mangos, pineapple, figs, corn, potatoes, rice, yams, nuts, cinnamon, vanilla are all grown in rainforests. Multiple life-saving medicines can be harvested from the rainforest. There is a wide range of animal life in tropical forests, many of which have bright colors and abstract designs to create fear of poisoning in potential predators. Most predators have camouflage colors. Rainforest animals have waxy coatings on their fur and feathers to repel water. Amphibians have extra slimy coating to protect them from the excess moisture. Monkeys, sloths, leafcutter ants, beetles, amphibians, jaguars, tigers, eagles, bears, and many bird species can be found living in the tropical rainforest. Many species of fungi, bacteria, and insects aid in nutrient recycling.

Originally tropical rainforests were global climate regulators because they absorbed large amounts of carbon dioxide during photosynthesis, and transpired as part of the water cycle, but with increased slash-and-burn destruction, the rainforest today produces more carbon dioxide than it absorbs. Slash-and-burn is a method of agriculture where tropical forest vegetation is cut down (slashed) and then burned to clear land for planting crops. The ashes from the burned vegetation temporarily enrich the soil with nutrients for farming, but agriculture quickly depletes the soil forcing the farmers to create new fields. Slash-and-burn results in poor air quality, excessive pollution, and irreparable damage by destroying many species before they have even been discovered. The illegal wildlife trade endangers countless animals through poaching and trading resulting in disrupted ecosystems and endangered organisms. Mining and resource extraction degrades and damages large areas of tropical rainforests.

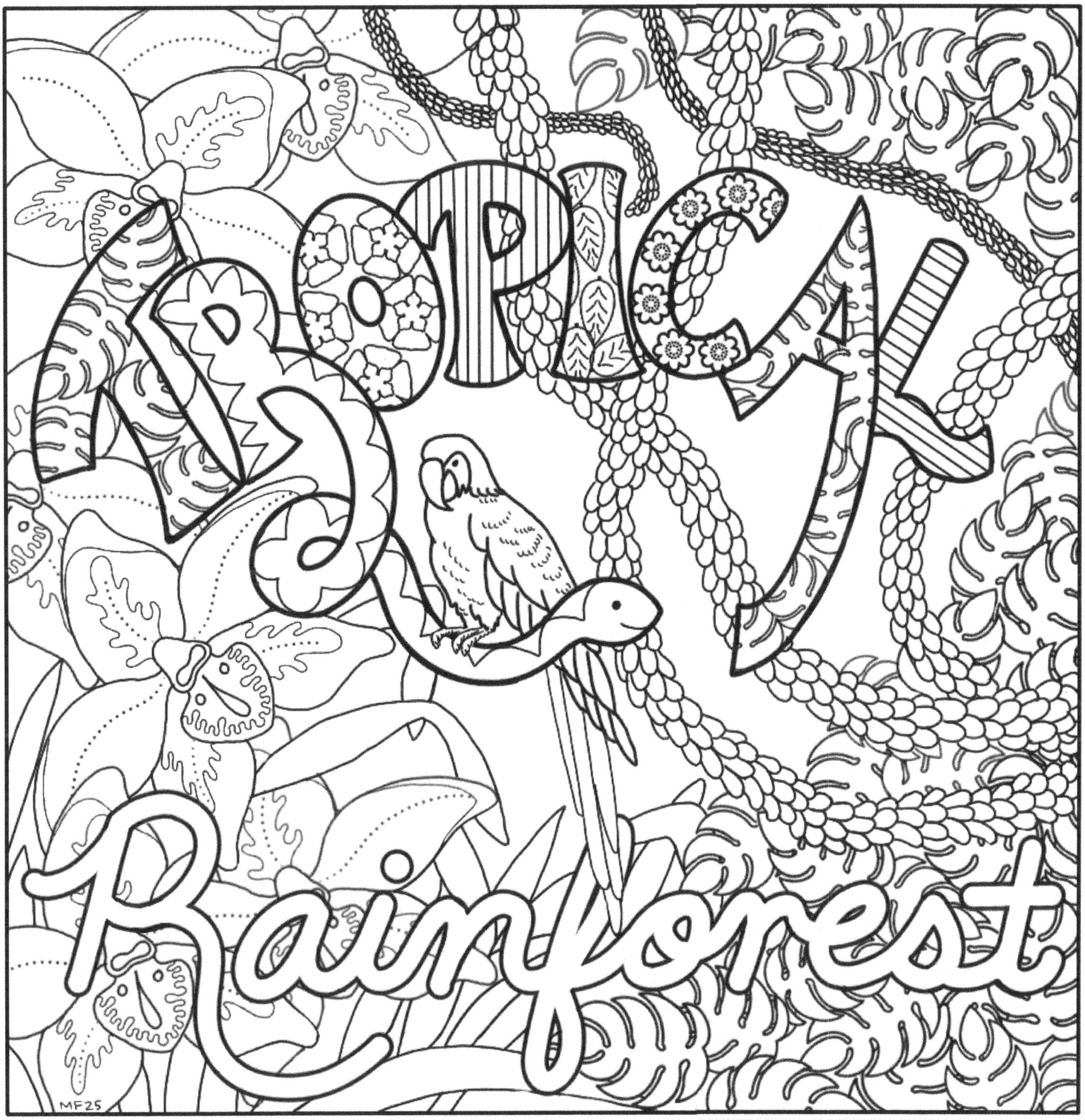

What are the defining characteristics of tropical rainforests?	What kinds of adaptations do animals have in the rainforest?
What kinds of foods grow in rainforests?	Why is slash-and-burn so detrimental?

Temperate Forest: A forest ecosystem found in regions with moderate temperatures and distinct seasons, including warm summers and cold dormant winters. It is characterized by deciduous trees that shed their leaves seasonally or a mix of deciduous and coniferous trees. The temperate forests are known for their rich biodiversity of plants and animals.

Abiotic elements in temperate forests are soil, water, temperature, sunlight, and wind. Temperate soil is fertile and rich in organic matter from fallen leaves, fruit and animal droppings. Many decomposers such as fungi, bacteria, snails, earthworms, and slugs helping to process the detritus (dead plant material) into soil. Temperate forests have moderate rainfall, ranging from 75-150 cm (30-60 inches) annually with possible snow. The temperature varies between -30°C to 30°C (-22°F and 86°F) across seasons. Temperate forests have seasonal variations in sunlight due to changing day lengths. Wind plays a major role in seed dispersal and affects temperature regulation.

Temperate forests contain high tree biodiversity. There are deciduous trees like oak, maple, birch and coniferous trees such as pine and fir. Many fruit trees can such as apple, apricots, peaches, persimmons, pears, plums, and cherries are native to temperate forests. Nut trees such as walnuts, chestnuts, pecans, almonds, filberts naturally grow in the forest. Temperate forests also contain shrubs, mosses, ferns, and multiple types of berries. Blackberries, strawberries, raspberries, elderberries, blueberries, and huckleberries can be found growing in the temperate forests. Fly agaric, chicken of the woods, chanterelle, turkey tail, and morels are common fungus species which help the recycling of forest nutrients. Underground truffles can be found using wild boars. In some cases, wild boars become dangerous invasive forest inhabitants.

Deer, rabbits, and squirrels are plentiful in the forest along with wolves, foxes, and owls. Omnivores such as bears, raccoons, and multiple bird species live in temperate forests. Bears and squirrels forage to gain weight in the summer and fall in order to hibernate through the winter. Most animals of the forest are various shades of brown and gray for ultimate camouflage. The animals have double coats for and thicker fat layers for protection against the cold winter. Some small reptiles like lizards, snakes, and turtles live in the forest along with amphibians, all of which hibernate in the cold.

Temperate forests face challenges from habitat fragmentation, urbanization, and exploitation of resources. Deforestation from the harvesting of wood and tree clearing for agriculture and urbanization significantly reduces the forest habitat leaving organisms without ample space to forage and find shelter. Expanding conservation policies, replanting trees, and teaching about sustainable resource collection can collectively protect temperate forests.

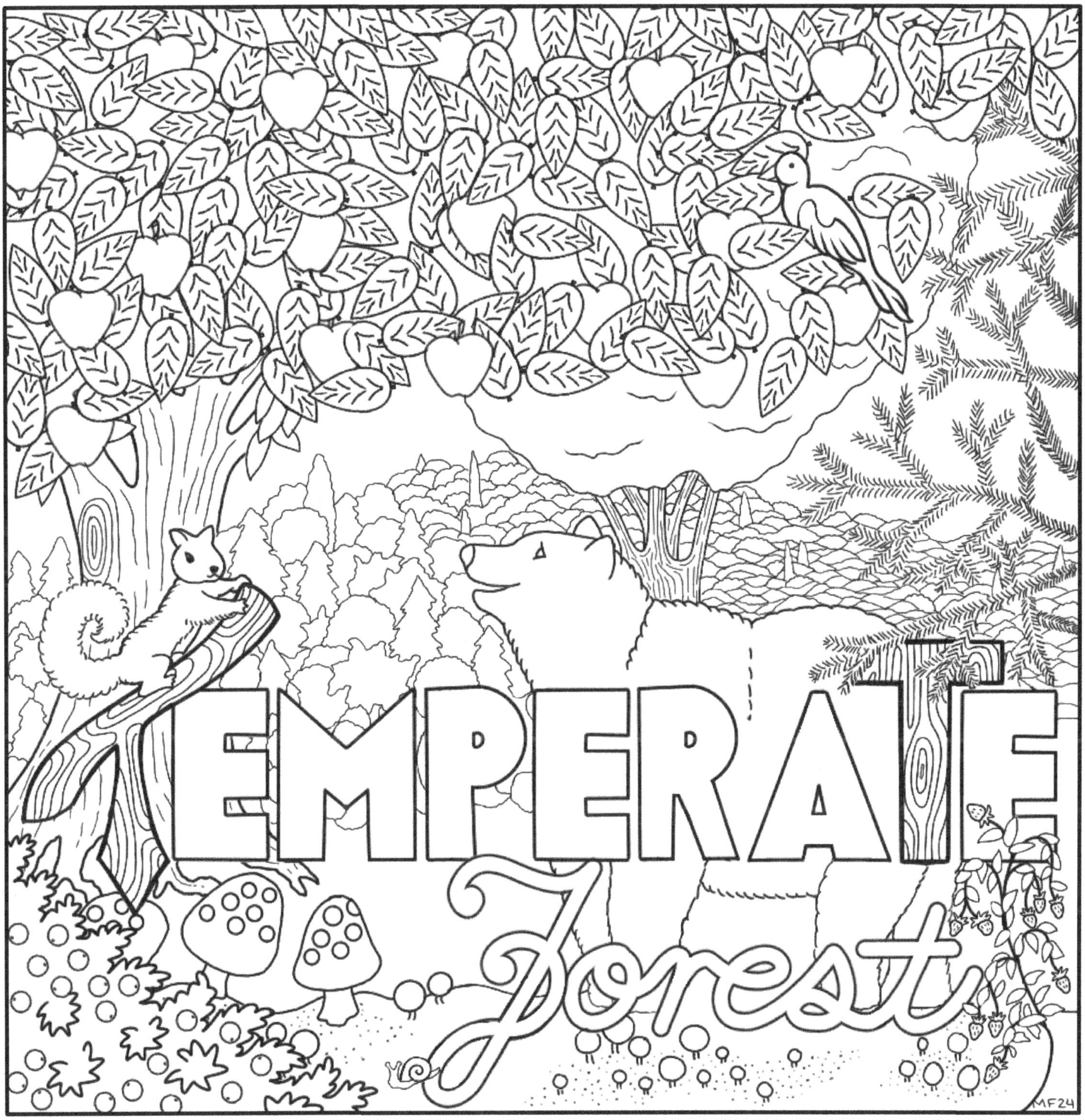

Describe the temperate forest climate.	What kinds of foods can be found in temperate forests?
Why do temperate forests have rich soil?	What kinds of ecological solutions can help protect and restore temperate forests?

Boreal Forest: A forest ecosystem, also known as the taiga, which is a vast, cold forest ecosystem found in the northern hemisphere. It is characterized by dense forests of coniferous trees and is the largest terrestrial biome on Earth. The boreal forest biome is dominated by coniferous trees such as spruce, fir, and pine, which are adapted to cold climates. These forests experience long, harsh winters and short, mild summers.

Boreal temperature varies from cool (21°C ,70°F) short summers, to very cold (-54°C, -65°F) long winters. Taiga soil is acidic, nutrient-poor soil with a thick layer of decomposing organic matter, some areas have permafrost layers where the ground remains frozen throughout the year. The precipitation of the boreal forest is moderate, ranging from 38-50 cm (15-20 inches) annually, mostly as snow. Because the taiga is only found in the northern hemisphere there are long daylight days in summer and short daylight days in winter.

Most of the trees in the boreal forest are evergreens. The evergreen needles resist freezing temperatures and water loss. Due to the high number of evergreens taigas are considered carbon sinks. All year the trees are actively photosynthesizing which absorbs carbon dioxide from the atmosphere. Various species of spruce, pine, alders, and even some willows grow in the taiga and provide food to the animals. The needles of the evergreens are rich in vitamin C while some have edible bark. Animals can forage bilberries, lingonberries, cloudberries, bearberries, and cranberries from the forest. There are also cattails, wild onion, chaga mushrooms, and morels. The forest floor is covered with sphagnum moss, club moss, and the symbiotic reindeer lichen. Plants adapt to the cold environment with waxy coatings. The shallow root systems allow them to get enough nutrients from the acidic soil, and in case of fires, the seeds are fire-resistant.

Animals of the boreal forest have special adaptations such as color changes with seasons, hibernation, burrowing, and foraging. Animals such as the snowshoe hare, ptarmigan, wolves, and foxes turn from brown colors to white and some gray in the winter to blend in with the snow. Wolverines and lynxes have darker fur patterns that help absorb heat from the sun. Bears and squirrels hibernate to survive the winter, while many bird species and caribou migrate south. Moose and deer forage to eat woody plants and bark. There are plenty of rodent species in the taiga; beavers, marmots, lemmings, porcupines, and voles. Wolves hunt in packs to eat moose and deer.

The taiga faces the danger from deforestation, and habitat fragmentation due to logging for timber, paper, and development. Climate change creates warming temperatures leading to permafrost thaw, increased wildfires, altered ecosystems, and the introduction of invasive species. Invasive earthworms release extra carbon making the soil even more acidic. Mountain pine beetle and emerald ash borer, spruce budworm, and gypsy moth are invasive insects that attack and kill various tree species by defoliation and damaging bark. Airborne pollutants and acid rain harm soil and plant life. Invasive plants such thistle, tansy, buckthorn, loosestrife, and canary grass outcompete native species.

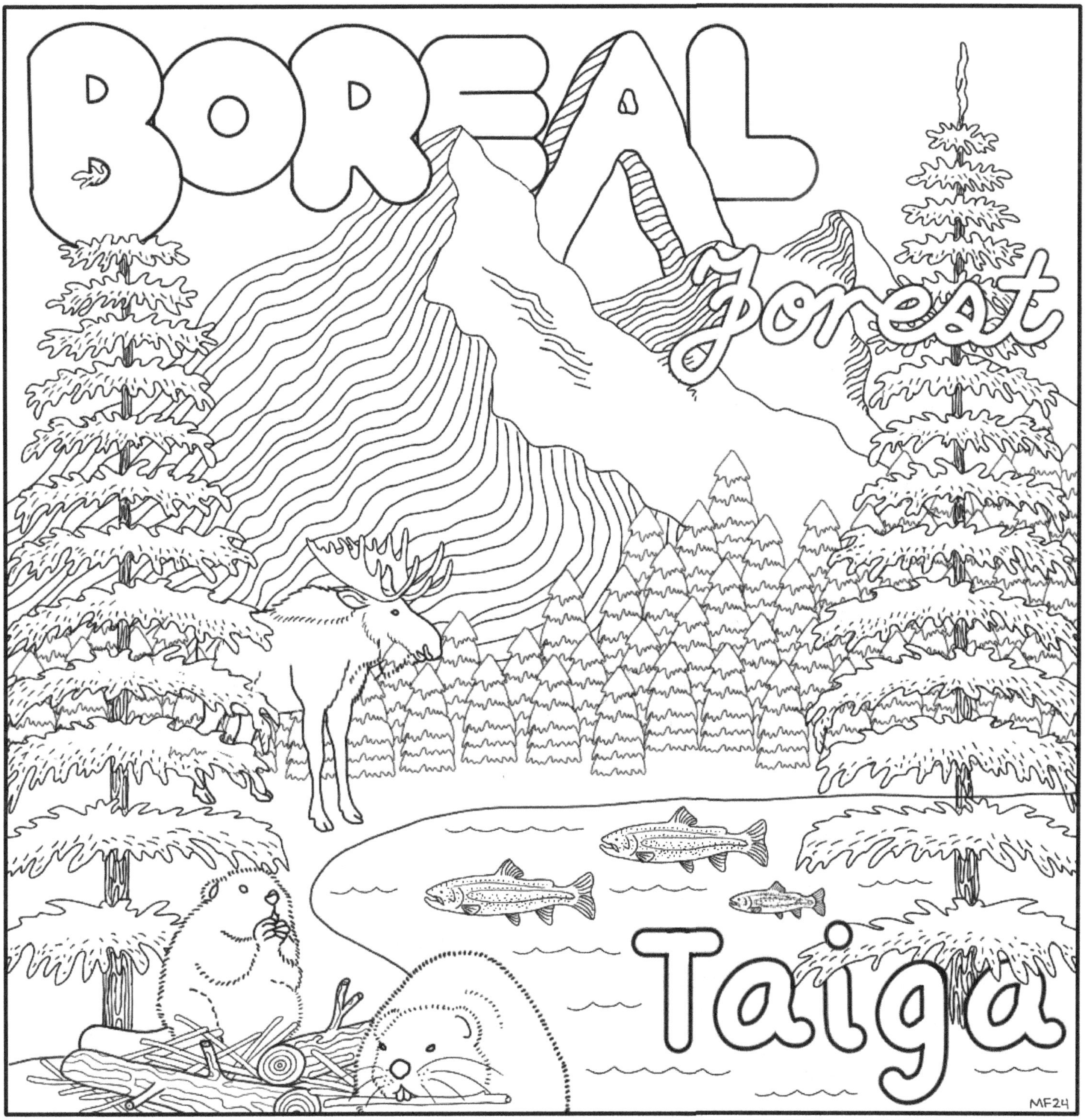

Which hemisphere are Boreal forests found in and what is another name for boreal forests?	Describe the plant adaptations of the boreal forest.
What are three abiotic factors that influence life in the boreal forest?	Describe the animal adaptations of the boreal forest.

Tundra Biome: A cold, treeless ecosystem found above the tree line in regions with harsh climates, minimal precipitation, and a short growing season. It is characterized by frozen ground (permafrost) and vegetation adapted to extreme cold and low nutrient availability. Most of the photosynthetic species are small shrubs, grasses, wildflowers and some lichen. The permafrost is a layer of soil that remains frozen year-round and stores carbon in the form of frozen plants. The permafrost prevents trees from growing because it is located below the active soil layer which is only 30 cm (12 in) thick. 30 cm is not enough space for trees to develop mature roots needed to be tall and healthy. The permafrost itself can be less than a meter thick to 1,500 meters thick (3-4,500 ft). Although there are few trees on the tundra, all the small ground covering plants are active with photosynthesis thereby categorizing tundras as carbon sinks.

Other important abiotic characteristics of the tundra are climate, precipitation, soil, sunlight, and wind. Extremely cold, dark winter temperatures as low as -34°C (-30°F) and short, cool, summers with temperatures averaging 3-12°C (37–54°F) and continuous daylight. Tundras receive low precipitation (snow) of 15-25 cm (6–10 in.) annually. The soil is nutrient poor and shallow. Strong winds, often exceeding 48 km/h (30 mph), contribute to the harsh conditions.

Tundra plants are small such as mosses, lichens, low shrubs, grasses, and flowering plants adapted to cold, short, growing seasons. Caribou, musk oxen, Arctic hares, lemmings, reindeer, Arctic foxes, wolves, and polar bears can be spotted in the tundra. Migratory birds such as geese, ducks, and terns spend time in the tundra during summer months. Decomposers like cold-adapted fungi, bacteria, and insects break down organic matter slowly due to the cold temperatures.

The tundras face problems with climate change, habitat destruction from over mining, excessive industrialization, resource depletion, pollution, and overgrazing. Global warming is melting the permafrost which can potentially release carbon dioxide, methane gas, ancient bacteria and viruses into the world. Efforts to reduce carbon dioxide levels in the atmosphere could help to slow global warming and reduce the speed at which the permafrost is melting.

TUNDRA

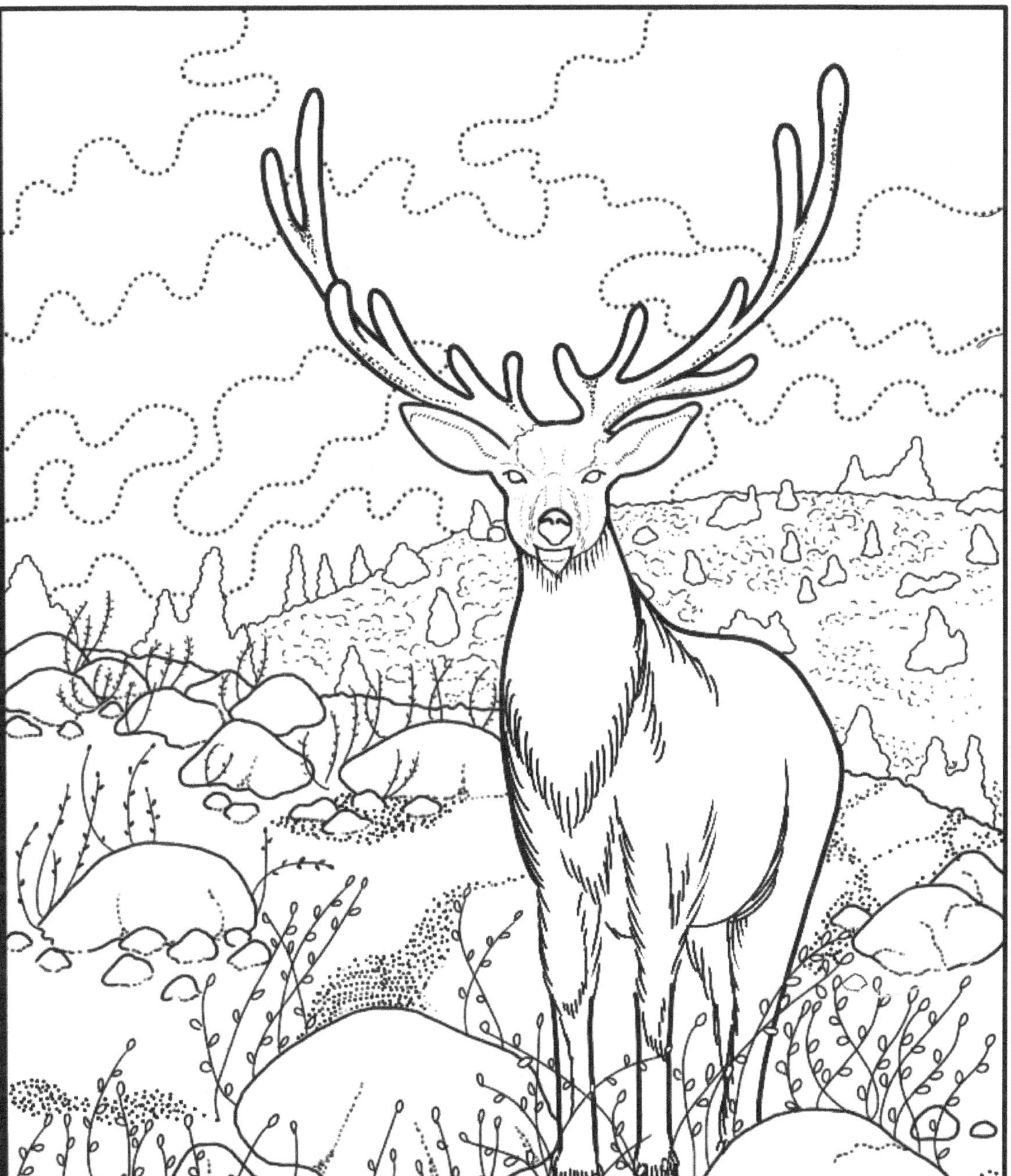

Describe tundras.

What is permafrost?

Using the context clues, what do you think "tree line" means?

What kinds of plants are found in the tundra?

Alpine Tundra: A high-altitude ecosystem found in mountainous regions worldwide above the tree line. The Alpine tundra starts at 3,350-3,505 meters (11,000-11,500 feet) above sea level. It is characterized by cold temperatures, strong winds, and a lack of trees due to the harsh climate and thin atmosphere. In some locations glaciers are present. The Alpine tundras are freshwater reservoirs support ½ of the world's population. The Alpine tundra has a short growing season and organisms with unique high altitude cold adaptations.

The Alpine tundra is cold year-round, with temperatures averaging below freezing during the winter and mild in the summer (2–16°C or 35–60°F). The grow season is short, lasting about 180 days or less resulting in small plants and limited vegetation. The soil is thin, rocky, and nutrient-poor, often well-drained but susceptible to erosion. The Alpine tundra receives mostly as snow, with annual precipitation ranging from 30–76 cm (12 to 30 inches). The humidity is low, due to the thin atmosphere at high altitudes. The tundra receives intense solar radiation due to high elevation, with fluctuating day lengths. Strong winds are common in the alpine tundra, further shaping the harsh conditions.

Alpine tundra plants are low-growing mosses, lichens, cushion plants, grasses, and wildflowers. Flowering plants, like the Edelweiss, have dense hairs on stems, leaves, and flowers to provide protection from the cold. Some plants have red pigment which absorb the sun's rays for heat. The cushion plants grow close to the ground and have long tap roots. altered and conserve nutrients it may take up to two years for flowering plants to develop flower buds. Pikas, marmots, mountain goats, and bighorn sheep are the alpine tundra's primary consumers. Alpine tundra carnivores are snow leopards, wolves, and birds of prey such as golden eagles. Most animals of the tundra have double fur coats with special fat layers for insulation. The hairs of the fur are hollow for more heat insulation while the hooves and paws are hairy to protect against snow, ice, and rocky terrain. Due to the high elevation, many animals have more red blood cells and larger lungs to help absorb more oxygen in the oxygen-poor atmosphere. Many animals hibernate during the cold months. The tundra has elevation-adapted species of beetles and butterflies. Decomposition is slow in the alpine tundra due to temperatures, but there are cold-adapted microorganisms and fungi present.

The Alpine tundra faces challenges with rising temperatures as glaciers melt faster, snow patterns are altered, and the habitat for plants and animals shift to even higher elevations. Tourism, skiing, and infrastructure development disturb fragile alpine ecosystems. Airborne pollutants and microplastics impact alpine tundra water sources and vegetation. Invasive species such as dandelion, common European earthworms, rainbow trout, and feral cats outcompete the native organisms. To protect the Alpine tundra, strict conservation efforts need to be put in place. It is vital that the Alpine freshwater sources are protected from pollution, contamination and human disruptions.

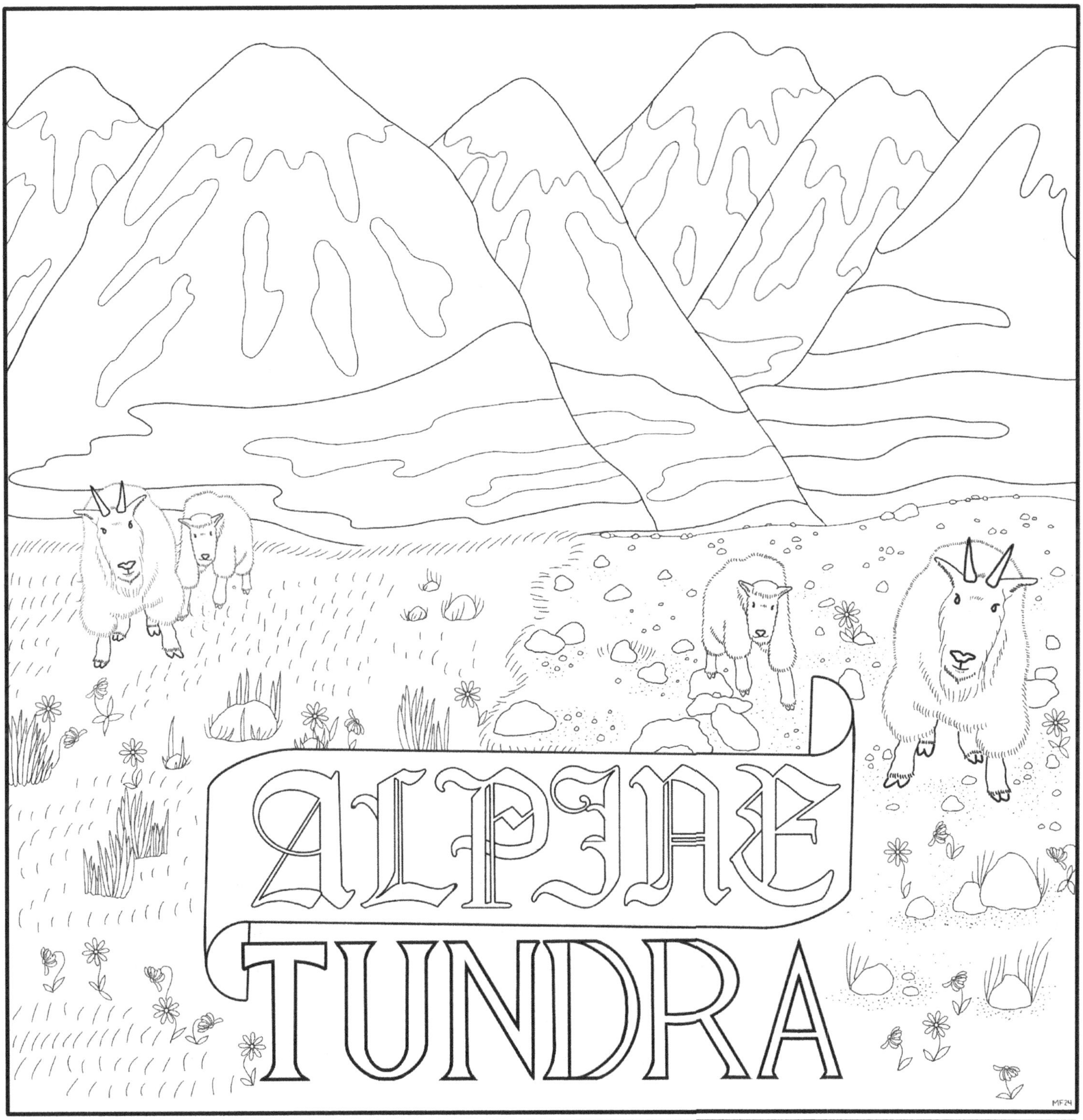

Where is the Alpine tundra located?	What kinds of adaptations do Alpine tundra animals have?
What kinds of adaptation do Alpine tundra plants have?	What are some things that negatively affect the Alpine tundra?

Arctic Tundra: A cold, treeless region found at high latitudes near the North Pole. It is characterized by permafrost, short growing seasons, and extreme environmental conditions like harsh winds.

Extremely cold climate, with winter temperatures as low as -50°C (-58°F) and summer temperatures rarely exceeding 10°C (50°F). The winters are long and harsh, while summers are short and cool. Arctic soil is thin, nutrient-poor, and often waterlogged due to permafrost. Arctic tundra days and nights are filled with continuous sunlight in summer and continuous darkness in winter. The precipitation of the arctic tundra is low like that of a desert with only about15-25 cm (6–10 inches) annually, mostly as snow. Due to the cold environment the small quantities of water the arctic receives is unable to evaporate. The Arctic tundra is constantly plagued by strong harsh winds.

The Arctic tundra has low-growing plants like mosses, lichens, grasses, and dwarf shrubs adapted to the cold and short growing seasons. Animals which make the Arctic their home include large mammals like caribou, musk oxen, and polar bears. There are also small animals like Arctic hares, lemmings, Arctic foxes, wolves, and snowy owls. Animals in the Arctic tundra have double coats of fur and multiple layers of feathers for warmth. The feet are wide with a lot of fur for protection against snow and ice. Polar bears have black skin under their fur to absorb the suns rays for heat. Most animals have small compact body sizes and ears to help retain heat. The Arctic tundra is near water edge, so multiple fish and aquatic mammals can also be spotted such as narwhals and walrus. The marine animals have extra layers of fat for insulation. Numerous migratory birds like geese, ducks, and terns visit during summer to breed. Decomposers such as fungi, bacteria, and soil microorganisms, can be found in the soil, although decomposition is extremely slow due to the cold climate.

The Arctic tundra faces catastrophic events from warming arctic temperatures. Ice flats and icebergs are melting at an alarming rate. The reduction in ice flats cause polar bears to starve. Without the ice in the ocean, polar bears have reduced opportunities to hunt and fish in the Arctic sea resulting in them dying from lack of food. The Arctic tundra needs stricter regulations on industrial activities like mining and oil extraction. Conservation efforts need to expand protections of native species and habitats through national parks and reserves. International agreements between countries must be made to limit activities that harm the Arctic including slowing down global warming by expanding the use of green energy.

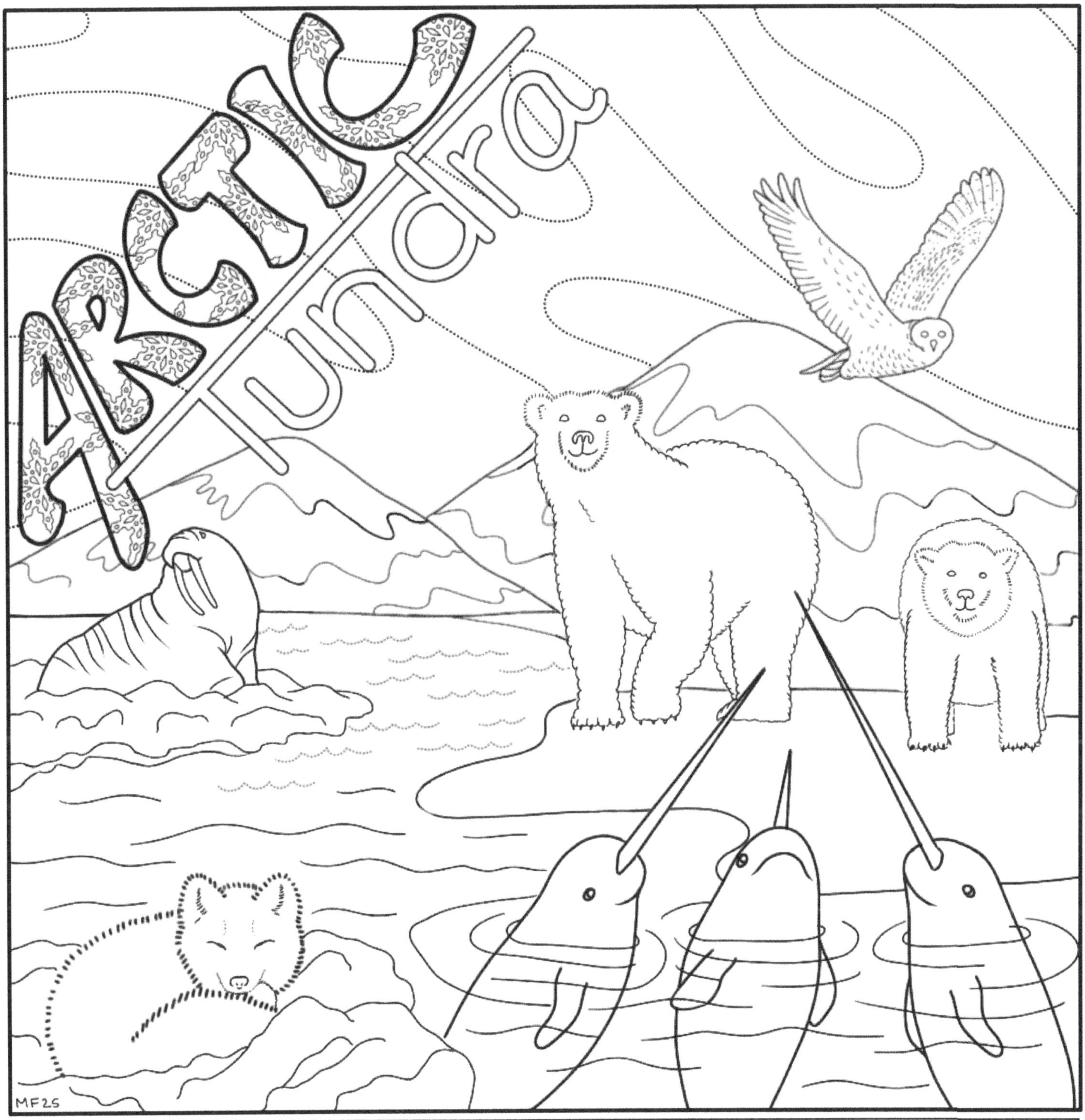

Describe the arctic tundra's climate.

Which animals live in the arctic tundra?

What kind of soil does the arctic tundra have?

List three things that could help preserve the Arctic tundra.

Antarctic Tundra: A cold, dry, treeless ecosystem located in Antarctica and surrounding sub-Antarctic islands. Unlike the Arctic tundra, it lacks large terrestrial mammals and has limited biodiversity due to extreme cold and isolation. The Antarctic tundra has extreme cold and dryness, with minimal vegetation growing on rocky, nutrient-poor soils. Ice and snow dominate most of the landscape, and life primarily exists in small, isolated patches near the coasts.

Winter temperatures of the Antarctic tundra can drop below -80°C (-112°F) and summer temperatures rarely exceed 5°C (41°F). The Antarctic tundra receives less than 5 cm (2 inches) of snow annually. The soil is rocky, dry, nutrient-poor, often covered by permafrost or ice. The tundra has experiences 24-hour sunlight in summer (polar days) and 24-hour darkness in winter (polar nights.) The wind is persistent, strong and katabatic (downward motion) which exacerbates the cold.

Antarctic flora are very small plant such as mosses, liverworts, algae, and lichens that are adapted to withstand the harsh cold conditions. These plants have fuzzy coatings to help retain heat and capture moisture. Some plants called cushion plants look like small green cushions laying close to the ground. Flowering plants like Antarctic hairgrass and pearlwort on sub-Antarctic islands are found in small numbers. There are only two tiny terrestrial microscopic organisms on the Antarctic land; tardigrades and nematodes. Most animals are marine-dependent, relying on the ocean for food. Penguins, and seals, have thick insulating fat. Seabirds and albatrosses have extra layers of down and feathers. Penguins have four layers of feathers and special nasal chambers to heat the air before it reaches their lungs. Within the ocean there are many cold-adapted fish and a few whales. The Antarctic waters are too cold for sharks. Due to the cold temperature the decomposers are very limited with very little nutrient recycling due to the lack of producers and terrestrial organisms.

Although the Antarctic tundra is cold and harsh, there are still threats that cause instability. Research stations, tourism, and fishing disturb the fragile ecosystem. Pollution from ships and stations affects land and water. Overfishing depletes key food sources for marine-dependent animals. Ozone depletion from industry and climate change results in increased UV radiation.

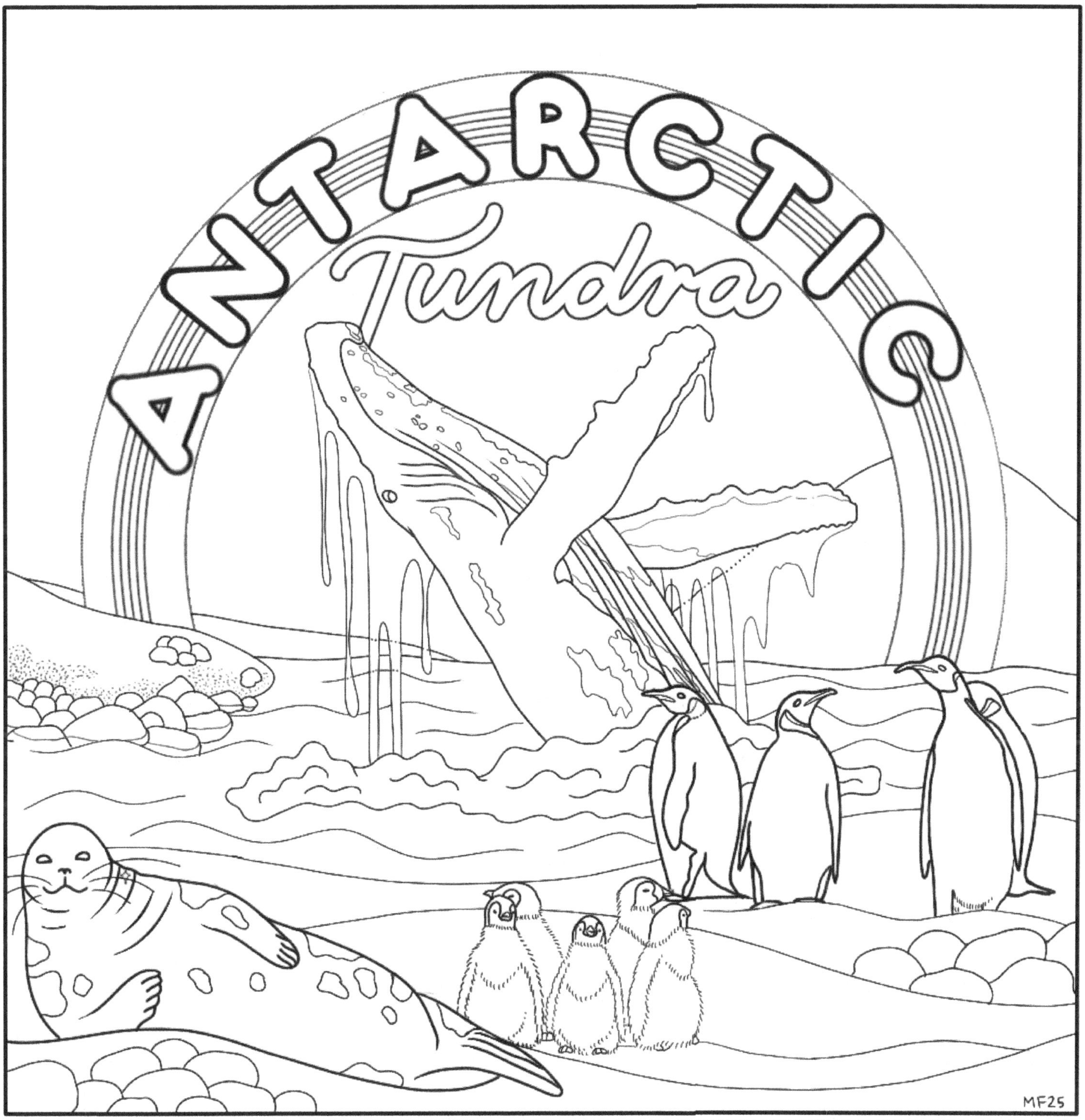

Where is the Antarctic tundra located?	What are polar days and polar nights?
What aren't there any large terrestrial mammals in the Antarctic tundra?	What is the main threat facing the marine-based animals?

Desert Biome: A dry, arid region characterized by low precipitation and extreme temperature variations. Deserts can be hot or cold, but all are defined by their lack of water, which significantly influences the plants, animals, and overall ecosystem. Deserts receive less than 25 cm (10 inches) of water annually. This scarcity of water shapes the desert's landscape, plant life, and animal adaptations. 20% of Earth's land is covered by deserts.

Temperatures in deserts can range from below freezing to 50°C (122°F) Desert soil is very sandy, rocky, and or gravelly, with low levels of organic material and nutrients. Desert sunlight is intense and abundant. Desert winds are strong, contributing to erosion, shifting sand dunes, animal adaptations, and behaviors. Hot deserts have high daytime temperatures, with significant cooling at night. Cold deserts have long, harsh winters with freezing temperatures and cool summers. Precipitation is extremely low, often sporadic and unpredictable. Due to low precipitation humidity is also low, leading to dry air. Evaporation of water leads to concentrations of minerals in desert lakes such as gypsum, sodium nitrate, sodium chloride, and borates.

Desert plants are adapted to live in dry harsh condition. Cacti, succulents, shrubs, and grasses dominate desert flora. Desert plants evolved adaptations like deep roots, water storage tissues, waxy coatings, and needles. Desert fauna include animals such as Kangaroo rats, camels, desert tortoises, foxes, snakes, scorpions, and birds of prey. Many animals are nocturnal to avoid daytime heat. Some animals estivate, spending hot and dry periods in a prolonged state of dormancy underground. The lack of water and humidity severely slows decomposition in the desert. Xerophilic fungus, bacteria, and insects decompose what they can. A lot of organic material simply dries out. It is not uncommon to find mummified dead animals and bones in the desert sand.

The desert biome is affected by climate change due to altered precipitation patterns and rising temperatures that can exacerbate droughts potentially turning semi-arid deserts into hot dry deserts. Desertification is a process where green ecosystems remain in drought so long, they become deserts.

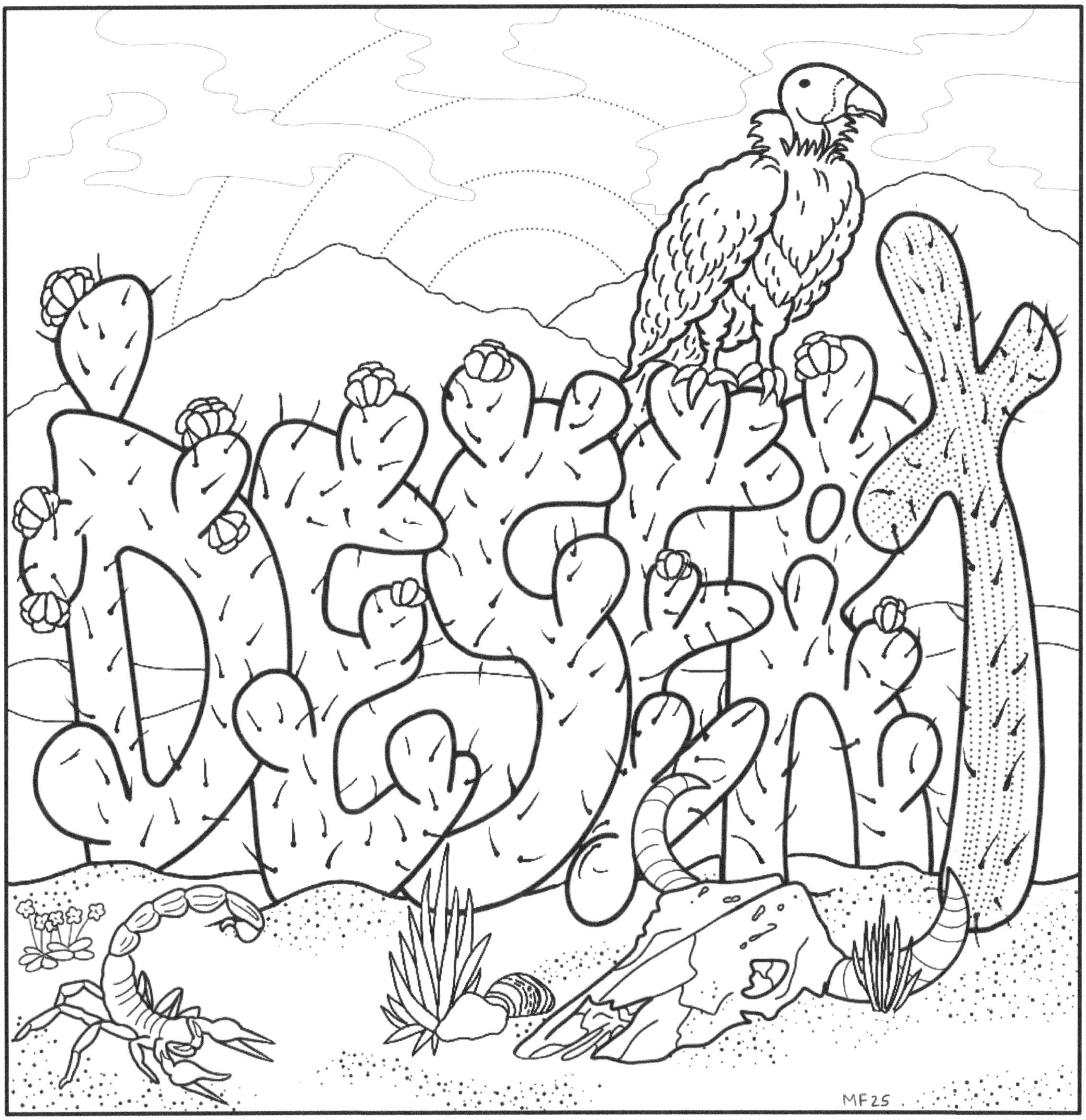

How much precipitation falls in the desert per year?	Why is decomposition slow in the desert?
What are animals doing when they estivate?	What is desertification?

Hot Dry Desert: The hot dry desert is a biome characterized by extremely high daytime temperatures, low precipitation, and arid conditions. These deserts are some of the most inhospitable places on Earth but are home to unique flora and fauna adapted to survive in the harsh environment. Temperatures can reach scorching highs during the day and plummet dramatically at night.

Hot dry desert temperature highs can exceed 49°C (120°F), while nighttime lows can approach freezing. Precipitation is sparse and irregular, often falling as brief, intense storms rarely ever exceeding 25 cm (10 in). Hot dry desert soil is primarily sandy with some gravel, low organic matter and poor water retention. The sunlight is intense and nearly constant, especially in summer. Strong winds contribute to erosion and shifting sands. Hot dry deserts are a dynamic landscape where sand dunes move, grow and are sometimes disappear. The desert's very low humidity leads to rapid evaporation of water.

Hot dry desert flora are both heat-tolerant and drought-resistant. A wide range of cacti, succulents, desert grasses, and shrubs live in the hot dry desert. They have adaptations such as waxy coatings, small leaves, and deep roots all designed to conserve water. The desert fauna such as Kangaroo rats, desert tortoises, and camels also have adaptations to survive the arid climate. Camels store water in their humps. Desert tortoises store large amounts of water in their bladders. Kangaroo rats have specialized kidneys, long nasal cavities and metabolic water production. Coyotes, foxes, snakes, lizards, and birds of prey make up the desert's predators. Most animals are nocturnal or crepuscular to avoid heat. Desert tortoises and toads estivate to survive the harsh climate. Decomposers such as insects, fungi, and bacteria that break down organic matter slowly due to low moisture.

Hot dry deserts face damage from oil and gas extraction. The drilling activities disturb habitats and pollute soil and water. Mining resources like uranium, gold, and salt damages the land and introduces toxic pollutants. In some places sand is being transported to other regions to rebuild beaches, make cement, and create water barriers leaving the desert land scared and barren. When sand is removed flora and fauna are negatively impacted.

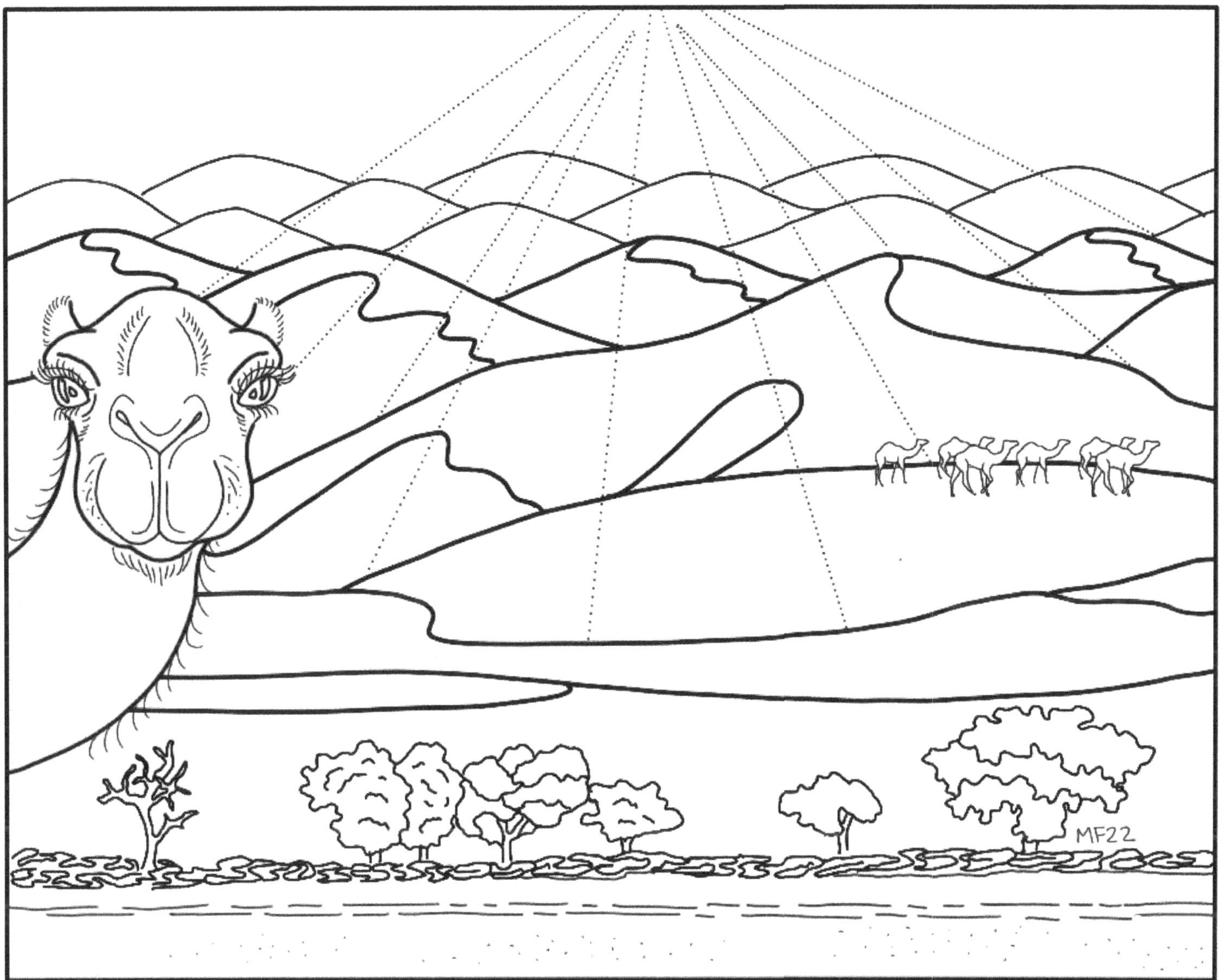

HOT DRY Desert

What is the temperature range of the hot dry desert?	What kinds of adaptations do animals have to survive the hot dry climate?
What are the abiotic elements of the hot dry desert?	Which human activities are negatively impacting the hot dry deserts?

Semi-arid Desert: A desert ecosystem that exists between arid deserts and more humid regions. It features slightly more precipitation than hot deserts, allowing for more vegetation and animal diversity. These deserts often have cold winters. The semi-arid desert has extreme temperature variations between seasons.

Semi-arid deserts experience hot summers often exceeding 38°C (100°F) and cold winters, with some temperatures dropping below freezing. Rainfall ranges from 25–50 cm (10–20 inches) annually. The soil is sandy, rocky, or gravelly, with low organic content but better water retention than hot dry desert soils. The sunlight is intense during the summer, with long daylight hours. Strong winds contribute to erosion and shape the landscape. Semi-arid deserts have low humidity but higher than arid deserts.

Semi-arid desert flora are shrubs, sagebrush, grasses, succulents, and drought-resistant plants like mesquite and cacti. These plants typically have adaptations for water storage such as specialized leaves, stems, roots, small leaves, needles, and thicker rough bark. Fauna of the semi-arid desert include both large and small mammals such as Jackrabbits, antelope, rodents, coyotes, and foxes. There are also many species of birds, and reptiles like lizards, Gila monsters, and snakes. Many animals have adapted to burrow or are crepuscular/nocturnal to avoid heat, a few amphibians and turtles even estivate during hot summers and hibernate during cold winters. Decomposers are drought and heat tolerant species of fungus, bacteria, and insects. Decomposition is slow in the semi-arid desert and often dead organisms are mummified by the sand and sun.

Semi-arid deserts are victim to urbanization and agriculture. When nonnative crops are planted, it causes a disruption in the environment, and excessive unsustainable amounts of water are used. Non-native plants also lead to rapid deterioration of the soil and unnecessary erosion. In semi-arid deserts it is important to plant native plant species that can tolerate the heat and drought conditions without wasting water while also stabilizing the soil with long roots. Native plants more provide food for insects, animals, and birds of the region.

To remediate human activities in the semi-arid desert, agriculture needs to focus on planting biodiverse native crops. The urban sprawl must be mitigated including landscaping. Xeriscaping while planting native species helps to reduce water consumption while supporting wildlife. Regulations against excessive mining of minerals, metals, and oil need to be established and followed.

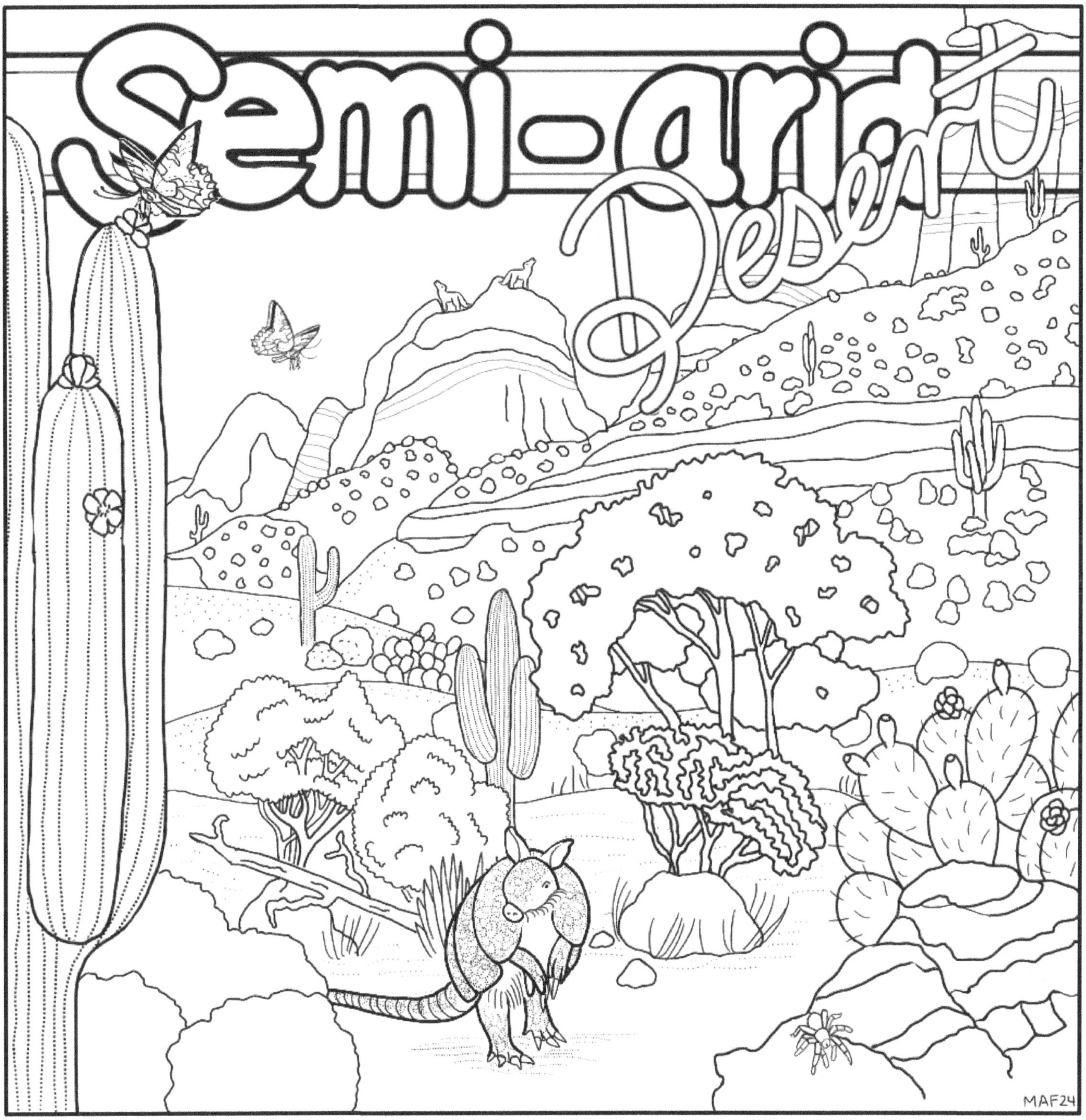

Describe the semi-arid desert ecosystem.

What kinds of adaptations do the fauna have?

How do non-native plants affect the semi-arid desert?

Why is it important to plant native plants in the semi-arid desert?

Coastal Desert: An ecosystem located along coastlines, characterized by arid conditions and influenced by cold ocean currents. Coastal deserts receive very little rainfall. The dry climate is combined with cool, foggy conditions, resulting from cold ocean currents and temperature inversions that prevent rain.

Coastal deserts have a mild but dry climate, with temperatures ranging from 10°C (50°F) to 24°C (74°F) during the day while nights are significantly cooler due to the absence of cloud cover. Precipitation is extremely low, typically less than 13 cm (5 in) per year. Dense fog provides a crucial source of moisture for the ecosystem. The soil is sandy or rocky, with poor nutrient content and low water retention. Coastal desert winds are often strong, contributing to the shaping of dunes and landscape erosion.

Coastal desert plants are drought-tolerant and salt-tolerant species like cacti, succulents, saltbushes, and lichens. Some flora have specialized leaf structures to harvest moisture from fog. Trichomes (fine hairs) on leaves trap moisture from fog and direct it toward the plant. Leaf surfaces are superhydrophilic which quickly condenses water from fog. Grooved or ridged leaves help collect fog and funnel condensed water toward the plant roots. Plants have shallow, wide-spreading roots designed to quickly absorb water from fog-drip before it evaporates or fog harvesting root systems.

The coastal desert has small animal species such as jackals, foxes, lizards, snakes, and small rodents like gerbils. Reptiles such as lizards and snakes are well adapted to arid conditions with their scaly skin which prevents water loss from evaporation, they urinate uric acid to conserve water and lack sweat glands to retain water. The reptile burrow to avoid high temperatures during the day and are more active at night. Even the sidewinding of snakes helps minimize the contact with hot sand. Some coastal species even have salt glands to secrete salts. Beetles and other insects have long limbs and run fast to reduce contact with hot sand. The Namib Desert beetle, use their bumpy exoskeleton to collect moisture from fog, which then trickles down into their mouths. Seabirds and desert-dwelling species like falcons soar in the skies above the coast desert sand dunes. Decomposition is limited due to low organic material, but bacteria, fungi, and some invertebrates do assist in the recycling of nutrients. Many dead organisms dry out in the salty sands.

The coastal desert is threatened by climate change because the changes in global climate alters ocean currents and fog patterns, disrupting the delicate balance of moisture organisms depend on. Coastal expansion for tourism and cities destroys natural habitats. Overfishing affects marine life that indirectly supports coastal desert by providing food to the sea birds. Plastic and chemical waste from human activity contaminates both land and marine species.

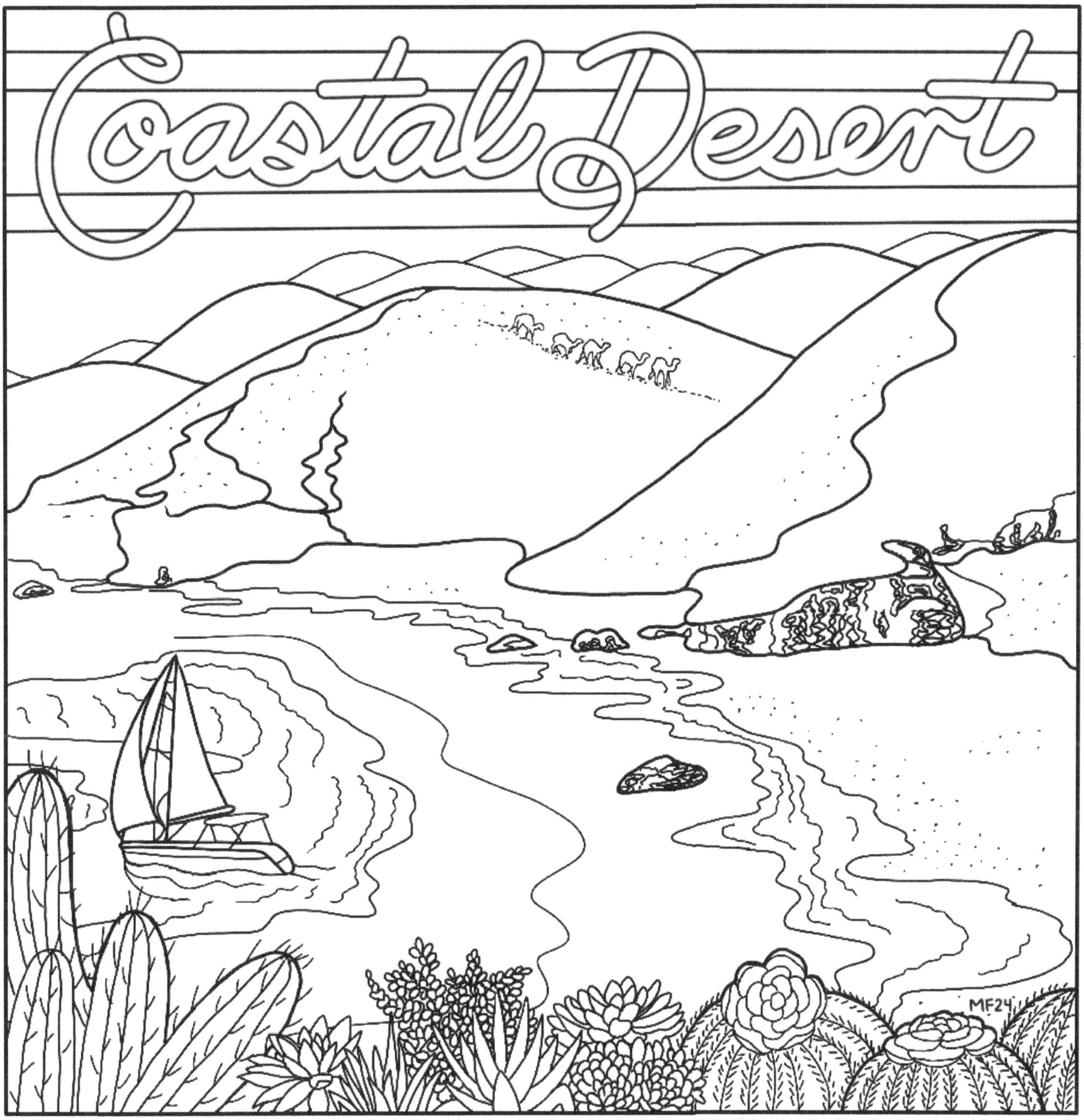

What is unique about the form of precipitation in coastal deserts?	What kind of adaptations do the fauna have to help them decrease contact with the hot sand?
How are the plants able to collect water from fog?	What adaptations do the animals have to help conserve water?

Cold Desert: A type of desert that experiences low temperatures and low precipitation. Unlike hot deserts, which are known for their scorching heat, cold deserts are characterized by their frigid temperatures and arid conditions. Organisms that live in the cold desert require adaptations for an arid cold environment.

The cold desert is characterized by low temperatures that drop well below freezing. Annual precipitation is less than 25 cm (10 in) per year, and most of it is snow. High freezing winds are common, which lead to soil erosion. The soil is very thin rocky and nutrient poor with minimal organic material. Winters are extremely cold, while summers can be cold, mild, or moderately warm. Water sources tend to be frozen rivers, glaciers, or seasonal meltwater. Depending on the geographic location the cold desert may have endless sunlit days in the summer with endless nights in the winter.

The cold desert has sparse vegetation. Plant life is limited to mosses, small shrubs, and drought-resistant grasses. Shrubs (e.g. sagebrush, rabbitbrush, saltbush) have small or narrow leaves for water conservation along with deep spreading root systems to absorb groundwater. Some plants have small wooly hairs and thick bark to protect against the cold. Cold desert animals such as snow leopards, camels, yaks, rodents, hares, and birds all have adaptations such as include thick fur, fat reserves, and burrowing behavior. Long slim legs help animals to move quickly and retain heat in their boxier bodies. Specialized paws and hoofs help the animals move in snow and rocky terrain. Some animals change colors with the season become white in the winter months to blend in with the snow. Birds of prey have keen eyesight to spot prey in the vast land. Reptiles undergo brumation, a slowed metabolic state to survive cold months. Reptiles also sunbathe to absorb heat and have dark skin and scales for maximum heat absorption. Microorganisms (bacteria and fungi) of the cold desert adapt to cold, dry conditions living in soil and under rocks.

The cold desert is threatened by climate change due to the rising temperatures that disrupt ecosystem leading to glacier and permafrost melting. As the climate warms up foreign species move in and compete with native species. Mining and resource extraction damage habitats and pollute the cold desert.

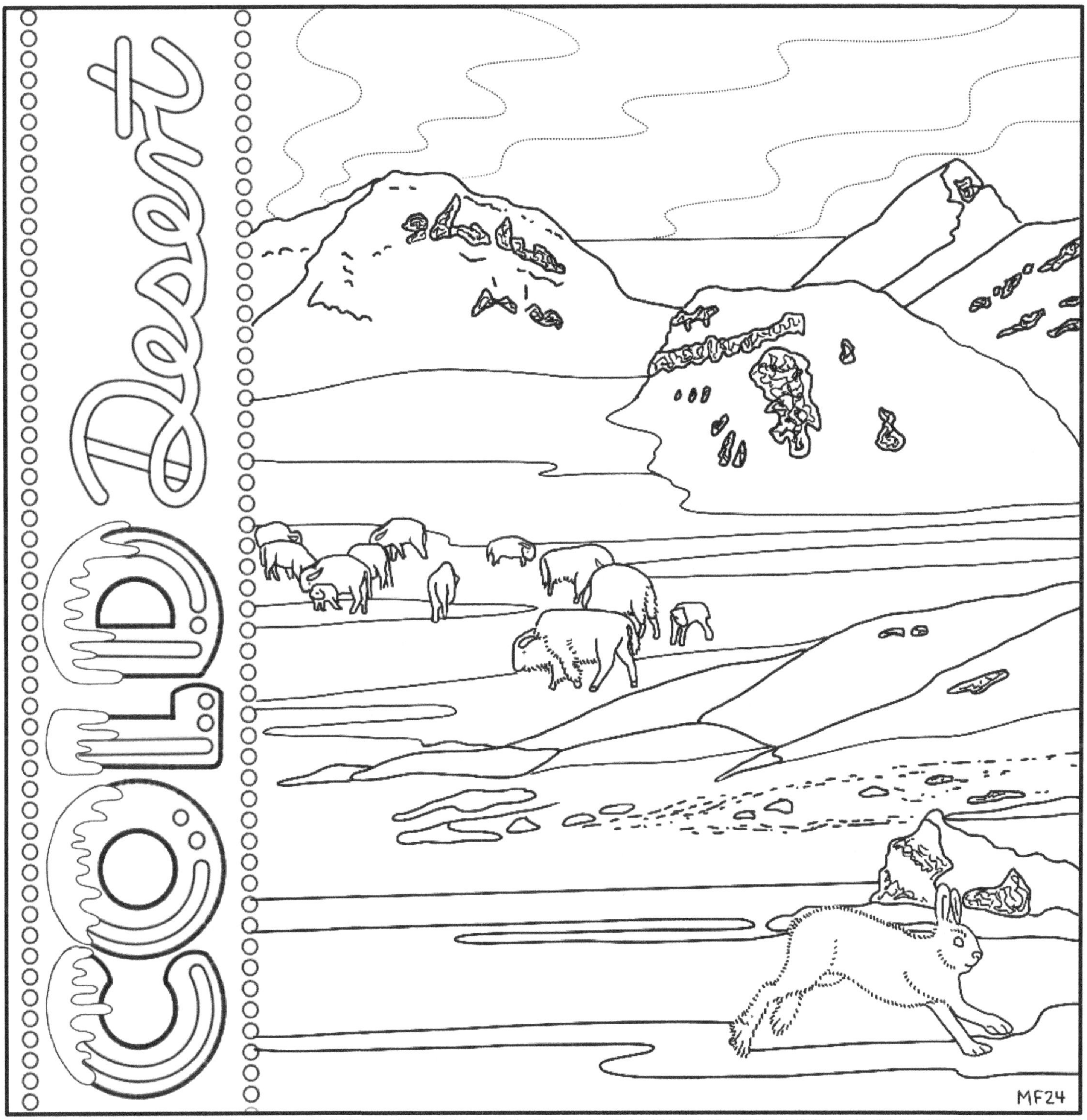

Describe the cold desert climate.

Why is the soil of the cold desert thin?

What kinds of adaptations do the plants have to survive in the ecosystem?

How does brumation help reptiles survive?

Caves: Natural underground or internal mountain spaces formed by geological processes, such as the dissolution of rock, volcanic activity, or erosion. They can vary in size from small chambers to vast networks of interconnected passages and are often found in limestone or other soluble rocks. The main characteristic of caves is their dark, enclosed environment with limited light, stable temperatures, and high humidity. These conditions create unique ecosystems that support specially adapted organisms.

Cave humidity is dependent on the climate of surface above. When there is more precipitation the cave will have more humidity, if the area above is in drought conditions, the cave will be too, although it may take longer to be evident than the surface. Most caves are composed of limestone, volcanic rock and have stalactites, stalagmites, and other formations from mineral deposits and erosion. Cave water may be dripping water, pooling water, rivers, and even some lakes. Temperatures are usually stable rarely fluctuating. Airflow in caves varies on the cave system, there can be areas with high airflow or very stagnant pockets with no oxygen.

Light levels in caves affects the lifeforms. The twilight zone is the cave entrance where natural light can still penetrate. Light levels are reduced compared to the outside but sufficient to support some photosynthetic life and animals called trogloxenes. Trogloxenes use caves for shelter or temporary purposes but must leave the cave to forage or survive. Barn owls and swifts can be found at the cave entrance roosting and nesting. Barn owls have sharp eyes and silent flights to help them hunt at night. Amphibians have moist skin to help absorb oxygen and a slowed metabolism to help survive the lack of food. Arthropods have longer legs and antenna to help detect motion and vibrations. Small mammals that live near the entrance have large round eyes, dense fur, and foraging capabilities. Bats use echolocation within their caves and outside to locate prey. Moss, liverworts, and ferns tolerate the high humidity, low light, and lack of soil. There are some species of algae, cyanobacteria, and lichen. Few flowering plants such as wild violet, begonias, and cave primrose grow in caves where light penetrates.

Beyond the twilight zone is the transition zone where light diminishes significantly. This is a transitional area between the entrance and the darkest parts of the cave, there is little to no photosynthesis in this area. Animals who live here are called troglophiles. They live and reproduce inside caves and rarely venture to the surface. Cave crickets and other cave-adapted insects have lost significant amounts of eyesight and even pigment. The deepest part of the cave is called aphotic zone where there is no light. Troglobites are animals that fully adapted to life in caves and cannot survive outside. Organisms in this zone rely on energy sources, such as organic material brought in from outside by other creatures, bat guano, or chemosynthetic bacteria. Most cave fish and various arthropod species are completely blind and must rely on enhanced senses which detect motion, vibration, taste, sound, and smell. They have very slow metabolisms which allow them to live on minimum food yet have long lives. Some cave salamander species can live up to 70 years.

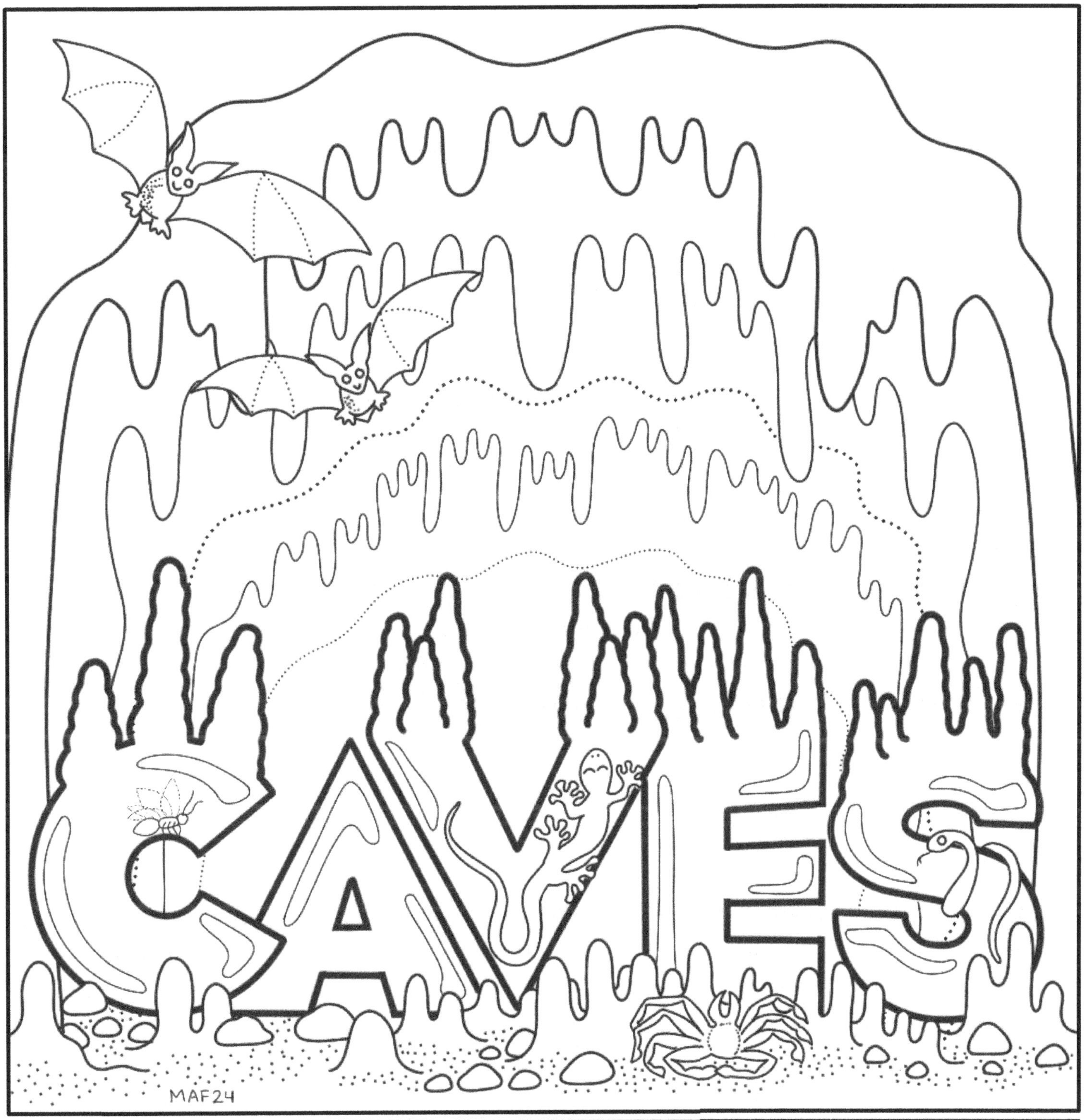

What are caves and what kinds of rock do caves form in? What is the deepest cave zone called and describe the organisms that live there?	Describe the climate inside of caves. Why do cave-dwelling animals depend on their other senses rather than their vision?

Urban Ecosystem: An urban ecosystem is a complex environment of interactions among living organisms, built environments, and the components within cities and towns. It includes parks, streets, buildings, people, plants, and animals, creating a unique blend of natural and human-made environments. The main characteristic of an urban ecosystem is its high degree of human influence. These ecosystems are defined by the integration of man-made environments, often marked by reduced biodiversity, altered natural processes and degradation of natural habitats due to urban development and sprawl.

Urban ecosystems create and affect abiotic environmental factors such as air quality, water resources, soil, temperature, and light levels. Urban ecosystems often have reduced air quality due to transportation, buildings which block air flow, pollution, and industry. The increased pollution and man-made surfaces such as buildings and streets increase temperatures. Inside cities and towns there will be higher air temperatures than a forest or desert due to buildings, sidewalks, and streets absorbing heat from the sun. Concrete, sidewalks, buildings, and streets also prevent water from precipitation from refilling natural water tables. Water resources such as rivers, lakes, reservoirs, and oceans are often subject to contamination from humans and industry. Vehicles introduce heavy metals and oil into ground water. Urban soil is often compacted, nutrient-poor and contaminated from pollution and lack of natural nutrient cycling. Light levels at night are higher than normal due to artificial lighting. For some animals and people, light pollution creates a disturbance to circadian rhythms (natural sleep/wake cycle of organisms).

Biotic factors in urban environments are the populations of people, and plants that were usually planted by the people. The plants that are introduced into urban spaces may not be native and will require more resources than native plants. The introduction of nonnative plants in urban environments creates food deserts for organisms native to the region. Foreign animals in urban environments may also be brought in by people introducing nonnative and invasive species. Sometimes native species become dependent on humans due to the disturbance of their natural food sources and having to make do with discarded food items from people. Racoons, cockroaches, mice, rats, and pigeons have evolved to follow and depend on human urbanization. They eat things that humans discard while making shelter in buildings and sewers. Urban ecosystems are at risk of taking more natural habitats to accommodate urban sprawl. Urban environments need to work on controlling pollution, habitat destruction, and negative affects of industrialization.

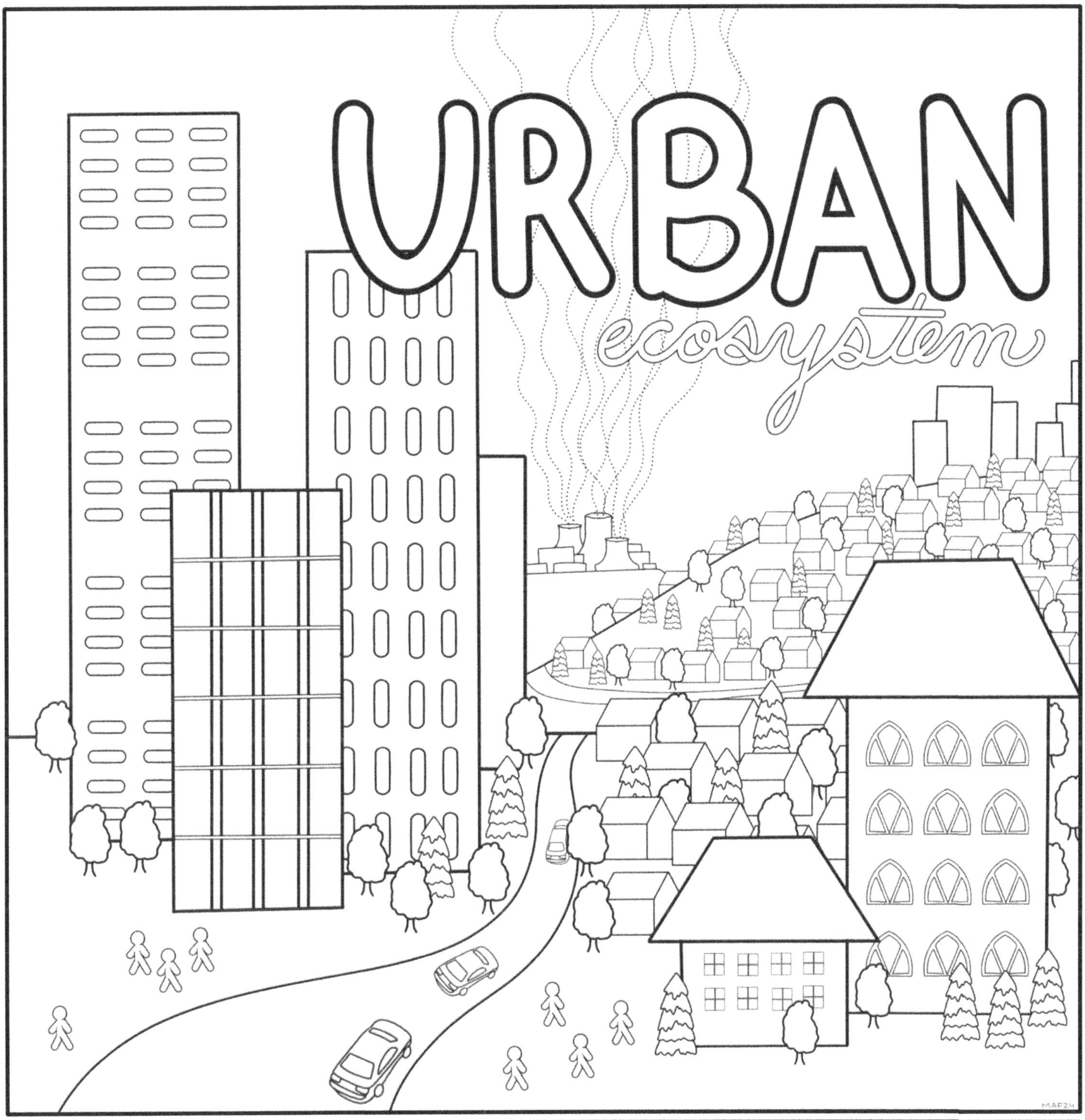

What is found in an urban ecosystem?	What kinds of animals are prevalent in urban ecosystems?
Why is light pollution problematic?	Describe the climate in urban ecosystems.

Made in United States
North Haven, CT
05 November 2026

21764042R00039